Problems and Solutions in
# Mathematical
# Olympiad

Secondary 1

# Problems and Solutions in
# Mathematical Olympiad

## Secondary 1

**Editors-in-Chief**
**Zun Shan** *Nanjing Normal University, China*
**Bin Xiong** *East China Normal University, China*

**Original Author**
**Zun Shan** *Nanjing Normal University, China*

**English Translator**
**Yi-Yang She** *Shanghai High School*

**Copy Editors**
**Ming Ni** *East China Normal University Press, China*
**Ling-Zhi Kong** *East China Normal University Press, China*
**Lei Rui** *East China Normal University Press, China*

**East China Normal University Press**

**World Scientific**

*Published by*

East China Normal University Press
3663 North Zhongshan Road
Shanghai 200062
China

and

World Scientific Publishing Co. Pte. Ltd.
5 Toh Tuck Link, Singapore 596224
*USA office:* 27 Warren Street, Suite 401-402, Hackensack, NJ 07601
*UK office:* 57 Shelton Street, Covent Garden, London WC2H 9HE

**Library of Congress Cataloging-in-Publication Data**
Names: Shan, Zun, author. | She, Yi-Yang, translator.
Title: Problems and solutions in Mathematical Olympiad : secondary 1 /
    original author, Zun Shan, Nanjing Normal University, China ;
    English translator, Yi-Yang She, Shanghai High School, China.
Description: Singapore ; Hackensack, NJ : World Scientfic, 2024.
Identifiers: LCCN 2024000547 | ISBN 9789811287206 (hardcover) |
    ISBN 9789811287428 (paperback) | ISBN 9789811287213 (ebook) |
    ISBN 9789811287220 (ebook other)
Subjects: LCSH: Mathematics--Problems, exercises, etc. | Mathematics--Competitions. |
    International Mathematical Olympiad.
Classification: LCC QA43 .S544 2024 | DDC 510.76--dc23/eng/20240316
LC record available at https://lccn.loc.gov/2024000547

**British Library Cataloguing-in-Publication Data**
A catalogue record for this book is available from the British Library.

For any available supplementary material, please visit
https://www.worldscientific.com/worldscibooks/10.1142/13701#t=suppl

Desk Editors: Nambirajan Karuppiah/Tan Rok Ting

Typeset by Stallion Press
Email: enquiries@stallionpress.com

# Editorial Board

## Board Members

**Hong-Bing Yu**     Ph.D. in Mathematics
Honorary Doctoral Supervisor
Professor, Suzhou University

**Zun Shan**     Ph.D. in Mathematics
Honorary Doctoral Supervisor
Chinese Team Leader, The 30th and 31st IMOs
Professor, Nanjing Normal University

**Shun-Qing Hang**     Mathematics Grand Grade Teacher
Senior Coach, The Chinese Mathematical Olympiad

**Xiong-Hui Zhao**     Ph.D. in Education
Deputy Director, Hunan Province Institute
of Educational Science

**Ming Ni**     Master Planner, *Mathematics Olympiad Book Series*
Director, Teaching Materials Branch,
East China Normal University Press

**Jun Ge**     Ph.D. in Education
Senior Coach,
The Chinese Mathematical Olympiad Principal,
High School Affiliated To Nanjing Normal University

**Bin Xiong**     Honorary Doctoral Supervisor
Director, Shanghai Key Laboratory of Pure
Mathematics and Mathematical Practice
Professor, East China Normal University
Member, The Chinese Mathematical
Olympiad Committee

# Preface

It is said that in many countries, especially the United States, children are afraid of mathematics and regard mathematics as an "unpopular subject." But in China, the situation is very different. Many children love mathematics, and their math scores are also very good. Indeed, mathematics is a subject that the Chinese are good at. If you see a few Chinese students in elementary and middle schools in the United States, then the top few in the class of mathematics are none other than them.

At the early stage of counting numbers, Chinese children already show their advantages.

Chinese people can express integers from 1 to 10 with one hand, whereas those in other countries would have to use two.

The Chinese have long had the concept of digits, and they use the most convenient decimal system (many countries still have the remnants of base 12 and base 60 systems).

Chinese characters are all single syllables, which are easy to recite. For example, the multiplication table can be quickly mastered by students, and even the slow learners know the concept of "three times seven equals twenty one." But for foreigners, as soon as they study multiplication, their heads get bigger. Believe it or not, you could try and memorize the multiplication table in English and then recite it, it is actually much harder to do so in English.

It takes the Chinese one or two minutes to memorize $\pi = 3.14159\cdots$ to the fifth decimal place. However, in order to recite these digits, the Russians wrote a poem. The first sentence contains three words and the second sentence contains one $\cdots$ To recite $\pi$, recite poetry first. In our opinion, this just simply asks for trouble, but they treat it as a magical way of memorization.

Application problems for the four arithmetic operations and their arithmetic solutions are also a major feature of Chinese mathematics. Since ancient times, the Chinese have compiled a lot of application questions, which has contact or close relations with reality and daily life. Their solutions are simple and elegant as well as smart and diverse, which helps increase students' interest in learning and enlighten students'. For example:

"There are one hundred monks and one hundred buns. One big monk eats three buns and three little monks eat one bun. How many big monks and how many little monks are there?"

Most foreigners can only solve equations, but Chinese have a variety of arithmetic solutions. As an example, one can turn each big monk into 9 little monks, and 100 buns indicate that there are 300 little monks, which contain 200 added little monks. As each big monk becomes a little monk 8 more little monks are created, so $200/8 = 25$ is the number of big monks, and naturally there are 75 little monks. Another way to solve the problem is to group a big monk and three little monks together, and so each person eats a bun on average, which is exactly equal to the overall average. Thus the big monks and the little monks are not more and less after being organized this way, that is, the number of the big monks is $100/(3 + 1) = 25$.

The Chinese are good at calculating, especially good at mental arithmetic. In ancient times, some people used their fingers to calculate (the so-called "counting by pinching fingers"). At the same time, China has long had computing devices such as counting chips and abaci. The latter can be said to be the prototype of computers.

In the introductory stage of mathematics — the study of arithmetic, our country has obvious advantages, so mathematics is often the subject that our smart children love.

Geometric reasoning was not well-developed in ancient China (but there were many books on the calculation of geometric figures in our country), and it was slightly inferior to the Greeks. However, the Chinese are good at learning from others. At present, the geometric level of middle school students in our country is far ahead of the rest of the world. Once a foreign education delegation came to a junior high school class in our country. They thought that the geometric content taught was too in-depth for students to comprehend, but after attending the class, they had to admit that the content was not only understood by Chinese students, but also well mastered.

The achievements of mathematics education in our country are remarkable. In international mathematics competitions, Chinese contestants have won numerous medals, which is the most powerful proof. Ever since our country officially sent a team to participate in the International Mathematical Olympiad in 1986, the Chinese team has won 14 team championships, which can be described as very impressive. Professor Shiing-Shen Chern, a famous contemporary mathematician, once admired this in particular. He said, "One thing to celebrate this year is that China won the first place in the international math competition $\cdots$ Last year it was also the first place." (Shiing-Shen Chern's speech, *How to Build China into a Mathematical Power*, at Cheng Kung University in Taiwan in October 1990)

Professor Chern also predicted: "China will become a mathematical power in the 21st century."

It is certainly not an easy task to become a mathematical power. It cannot be achieved overnight. It requires unremitting efforts. The purpose of this series of books is: (1) To further popularize the knowledge of mathematics, to make mathematics be loved by more young people, and to help them achieve good results; (2) To enable students who love mathematics to get better development and learn more knowledge and methods through the series of books.

"The important things in the world must be done in detail." We hope and believe that the publication of this series of books will play a role in making our country a mathematical power. This series was first published in 2000. According to the requirements of the curriculum reform, each volume is revised to different degrees.

Well-known mathematician, academician of the Chinese Academy of Sciences, and former chairman of the Chinese Mathematical Olympiad Professor Yuan Wang, served as a consultant to this series of books and wrote inscriptions for young math enthusiasts. We express our heartfelt thanks. We would also like to thank East China Normal University Press, and in particular Mr. Ming Ni and Mr. Lingzhi Kong. Without them, this series of books would not have been possible.

Zun Shan and Bin Xiong
May 2018

# Contents

# Chapter 1

# Ingenious Computation of Rational Numbers

The computation of rational numbers is similar to the computation of positive numbers. We should pay attention to the applications of operation properties (such as the commutative property of addition, the associative property of addition, the commutative property of multiplication, the associative property of multiplication, and the distributive property of multiplication) and skills in order to make the computation easy and convenient.

**Example 1.** Compute

$$48\frac{3}{5} - 18\frac{1}{4} - 1\frac{1}{3} + 0.25 + 3\frac{2}{3} - 2\frac{1}{3} - 30\frac{3}{5}.$$

**Solution**

$$48\frac{3}{5} - 18\frac{1}{4} - 1\frac{1}{3} + 0.25 + 3\frac{2}{3} - 2\frac{1}{3} - 30\frac{3}{5}$$

$$= \left(-1\frac{1}{3} + 3\frac{2}{3} - 2\frac{1}{3}\right) + \left(48\frac{3}{5} - 30\frac{3}{5}\right) + \left(-18\frac{1}{4} + \frac{1}{4}\right)$$

$$= \left(-\frac{1}{3} + \frac{2}{3} - \frac{1}{3}\right) + (48 - 30 - 18) + \left(\frac{3}{5} - \frac{3}{5}\right) + \left(-\frac{1}{4} + \frac{1}{4}\right)$$

$$= 0 + 0 + 0 + 0$$

$$= 0.$$

Remark: During addition and subtraction, we should pay attention to the application of the commutative and associative properties. We can change the order of operations in the computation, making the intermediate result an integer or even "cancel out." "Cancel out" means the sum of two oppo-site numbers is zero, such as the sum of $\frac{3}{5}$ and $-\frac{3}{5}$ and the sum of $-\frac{1}{4}$ and $\frac{1}{4}$ in this example. But during the computation, be careful of the signs and ensure that there are no mistakes. For example, $-\frac{1}{3}$ and $-\frac{1}{3}$ cannot cancel out, but their sum can, giving $\frac{2}{3}$.

**Example 2.** Compute

$$-2.5 \div 0.75 \times \left(-\frac{1}{5}\right) \times \left(-1\frac{3}{4}\right) \div (-1.4) \times \left(-\frac{3}{5}\right) \times \frac{2}{3}.$$

**Solution**

$$-2.5 \div 0.75 \times \left(-\frac{1}{5}\right) \times \left(-1\frac{3}{4}\right) \div (-1.4) \times \left(-\frac{3}{5}\right) \times \frac{2}{3}$$

$$= -\frac{5}{2} \div \frac{3}{4} \times \frac{1}{5} \times \frac{7}{4} \div \frac{14}{10} \times \frac{3}{5} \times \frac{2}{3}$$

$$= -\frac{10}{3} \times \frac{1}{5} \times \frac{7}{4} \times \frac{10}{14} \times \frac{3}{5} \times \frac{2}{3}$$

$$= -\frac{1}{3}.$$

Remark: During multiplication and division, we should pay attention to the sign of the result: the multiplication of an odd number of negative numbers is a negative number; the multiplication of an even number of negative numbers is a positive number. Usually, we convert decimals to fractions, mixed fractions to improper fractions, and division to multiplication. Simplify the fraction first, and make the computation as easy as possible.

In addition, we should memorize some common results, such as $0.125 = \frac{1}{8}$, $0.375 = \frac{3}{8}$, and $0.75 = \frac{3}{4}$.

**Example 3.** Compute
$$\frac{1 \times 2 \times 3 + 2 \times 4 \times 6 + 7 \times 14 \times 21}{1 \times 3 \times 5 + 2 \times 6 \times 10 + 7 \times 21 \times 35}.$$

**Solution**

$$\frac{1 \times 2 \times 3 + 2 \times 4 \times 6 + 7 \times 14 \times 21}{1 \times 3 \times 5 + 2 \times 6 \times 10 + 7 \times 21 \times 35}$$

$$= \frac{1 \times 2 \times 3 \times (1 + 2 \times 2 \times 2 + 7 \times 7 \times 7)}{1 \times 3 \times 5 \times (1 + 2 \times 2 \times 2 + 7 \times 7 \times 7)}$$

$$= \frac{2}{5}.$$

Remark: During computation, we should pay attention to the application of the distributive property. If the numbers added have a common factor, we can first extract the common factor, then compute the parts with their common factor removed. In this example, the numerator has a common factor of $1 \times 2 \times 3$ and the denominator has a common factor of $1 \times 3 \times 5$; therefore, we can extract them and then simplify in order to make the computation easier. Moreover, if the common factor is a negative number, after extracting the common factor, the sign of each remaining part will change.

**Example 4.** Compute

$$\frac{1}{2} + \frac{1}{4} + \frac{1}{8} + \frac{1}{16} + \frac{1}{32} + \frac{1}{64}.$$

**Solution**

$$\frac{1}{2} + \frac{1}{4} + \frac{1}{8} + \frac{1}{16} + \frac{1}{32} + \frac{1}{64}$$

$$= \frac{1}{2} + \frac{1}{4} + \frac{1}{8} + \frac{1}{16} + \frac{1}{32} + \left( \frac{1}{64} + \frac{1}{64} \right) - \frac{1}{64}$$

$$= \frac{1}{2} + \frac{1}{4} + \frac{1}{8} + \frac{1}{16} + \left( \frac{1}{32} + \frac{1}{32} \right) - \frac{1}{64}$$

$$= \frac{1}{2} + \frac{1}{4} + \frac{1}{8} + \left( \frac{1}{16} + \frac{1}{16} \right) - \frac{1}{64}$$

$$= \cdots$$

$$= \left( \frac{1}{2} + \frac{1}{2} \right) - \frac{1}{64}$$

$$= 1 - \frac{1}{64}$$

$$= \frac{63}{64}.$$

Remark: In this example, we can find a feature of the summation formula: every succeeding term is half of the preceding term. So, if we add the next term by itself, the result is simply the preceding term. Therefore, we ingeniously add $\frac{1}{64}$ to the formula and compute. Of course, since we add $\frac{1}{64}$ to the formula, we must also subtract $\frac{1}{64}$ at the end.

**Example 5.** Compute

(1) $1 + 2 + 3 + 4 + \cdots + 2007 + 2008$;

(2) $1 - 2 + 3 - 4 + \cdots + 2007 - 2008$.

**Solution**

(1) Let $S = 1 + 2 + \cdots + 2007 + 2008$. Then, $S = 2008 + 2007 + \cdots + 2 + 1$. After adding these two equations, we have

$$2S = (1 + 2008) + (2 + 2007) + \cdots + (2007 + 2) + (2008 + 1)$$

$$= \underbrace{2009 + 2009 + \cdots + 2009 + 2009}_{2008}$$

$$= 2009 \times 2008.$$

Then, we get $S = \frac{2009 \times 2008}{2} = 2017036$. To conclude, the original formula is equal to $2017036$.

(2)

$$1 - 2 + 3 - 4 + \cdots + 2007 - 2008$$

$$= (1 - 2) + (3 - 4) + \cdots + (2007 - 2008)$$

$$= \underbrace{(-1) + (-1) + \cdots + (-1)}_{1004}$$

$$= -1004.$$

**Remark:** The feature of problem (1) is that the difference between any two consecutive terms is the same value. We call such a sequence an arithmetic sequence. In other words, if a sequence $a_1, a_2, \ldots, a_n$ satisfies the condition that $a_{i+1} - a_i = d$ holds for all $i = 1, 2, \ldots, n - 1$, then we call such a sequence an arithmetic sequence. Here, $a_1$ is called the first term, $a_n$ is called the last term, and $d$ is called the common difference. The formula for computing $a_1 + a_2 + \cdots + a_n$ is

$$\text{Sum} = \frac{(\text{the first term} + \text{the last term}) \times \text{the number of terms}}{2}.$$

Now, we can solve problem (1) using this formula.

Sometimes, the number of terms is not shown in the problem. We can compute it using this formula:

$$\text{the number of terms} = \frac{\text{the last term} - \text{the first term}}{\text{the common difference}} + 1.$$

For problem (2), we combine the two adjacent terms to simplify the computation. This is determined by using the feature of this problem. When solving problems, do not rush to the computation; rather, we should first observe the feature of the problem, and then start solving it from the feature, which can make half the effort, twice the result.

**Example 6.** Compute

$$\frac{1}{1 \times 2} + \frac{1}{2 \times 3} + \frac{1}{3 \times 4} + \cdots + \frac{1}{1999 \times 2000}.$$

**Solution**

$$\frac{1}{1 \times 2} + \frac{1}{2 \times 3} + \frac{1}{3 \times 4} + \cdots + \frac{1}{1999 \times 2000}$$

$$= \left(1 - \frac{1}{2}\right) + \left(\frac{1}{2} - \frac{1}{3}\right) + \left(\frac{1}{3} - \frac{1}{4}\right) + \cdots + \left(\frac{1}{1999} - \frac{1}{2000}\right)$$

$$= 1 + \left(-\frac{1}{2} + \frac{1}{2}\right) + \left(-\frac{1}{3} + \frac{1}{3}\right) + \cdots + \left(-\frac{1}{1999} + \frac{1}{1999}\right) - \frac{1}{2000}$$

$$= 1 - \frac{1}{2000}$$

$$= \frac{1999}{2000}.$$

Remark: During addition and subtraction, by using the feature of numbers, we can split every number into two, after which some of them can cancel out. We call this the "split method." In this example, we split $\frac{1}{n \times (n+1)}$ into $\frac{1}{n} - \frac{1}{n+1}$, which is

$$\frac{1}{n \times (n+1)} = \frac{1}{n} - \frac{1}{n+1}.$$

There are some other ways of splitting:

(1) $\frac{d}{n \times (n+d)} = \frac{1}{n} - \frac{1}{n+d}$ or $\frac{1}{n \times (n+d)} = \frac{1}{d}(\frac{1}{n} - \frac{1}{n+d})$. Usually, this is applied when the factors of the denominators form an arithmetic sequence with common difference $d$.

(2) $\frac{1}{n \times (n+1) \times (n+2)} = \frac{1}{2} \times [\frac{1}{n \times (n+1)} - \frac{1}{(n+1) \times (n+2)}]$.

**Example 7.** Adding $\frac{1}{2}$ of 2002 to 2002, we get the first number. Then, adding $\frac{1}{3}$ of the first number to itself, we get the second number. Then, adding $\frac{1}{4}$ of the second number to itself, we get the third number, and so on in the same way until we add $\frac{1}{2002}$ of the current number to itself to get the last number. What is the last number?

**Solution** Since 2002 plus $\frac{1}{2}$ of it is $2002 \times (1 + \frac{1}{2})$, and then we add $\frac{1}{3}$ of this number to get $2002 \times (1 + \frac{1}{2}) \times (1 + \frac{1}{3})$, and so on, and the last number is equal to

$$2002 \times \left(1 + \frac{1}{2}\right) \times \left(1 + \frac{1}{3}\right) \times \cdots \times \left(1 + \frac{1}{2002}\right)$$

$$= 2002 \times \frac{3}{2} \times \frac{4}{3} \times \cdots \times \frac{2003}{2002}$$

$$= 2002 \times \frac{2003}{2}$$

$$= 2005003.$$

Remark: During multiplication, canceling the number that appears both in the numerator and denominator can greatly simplify the computation.

## Reading

---

### Rational Numbers

If you are new to rational numbers, then you might ask: Why are numbers of the form $\frac{m}{n}$ (where $m$ and $n$ are integers, $n \neq 0$) called rational numbers? Since there are rational numbers, are there irrational numbers?

Usually, it makes sense to give a thing a name. For example, the word "negative" of a negative number has the meaning of debt, and its meaning is exactly the opposite of the word "positive" of a positive number. The reason why a rational number is called so is unreasonable. It stems from a mistake in translation.

In the 19th century, when Western science was introduced to China, the Chinese mathematician Li Shanlan (1811–1882) translated rational function and irrational function into proportional and non-proportional formulas, respectively, when translating British mathematician De Morgan's *Algebra*. The understanding of these two names was completely correct, and the translated names were also correct because "proportional" refers to ratio. But more than 10 years later, when another mathematician, Hua Hengfang (1833–1902), translated Wallace's *Algebra*, he mistranslated rational and irrational as reasonable and unreasonable, respectively, which did not match the original meaning, but they became widely spread, even into Japan. Now, both China and Japan use these incorrect translations in education and academia.

As a rational number is a number that can be represented as $\frac{m}{n}$ (where $m$ and $n$ are integers, $n \neq 0$), a number that cannot be represented as $\frac{m}{n}$ is an irrational number. Such numbers exist. For example, the ancient Greeks discovered that the side length of a square is an irrational number if its area is 2.

## Exercises

Compute:

1. $31\frac{2}{7} - 22\frac{6}{13} + 4\frac{5}{7} + 11\frac{6}{13}$.

2. $5\frac{6}{11} - 3.125 - 7\frac{4}{7} - 3\frac{4}{11} + 8\frac{1}{8} - 3\frac{6}{7} - 2\frac{2}{11} + 6\frac{3}{7}$.

3. $-\frac{7}{11} \div 2.5 \times (-0.75) \div (-1\frac{2}{5}) \div \frac{3}{11} \times (-\frac{8}{13})$.

4. $3.825 \times \frac{1}{4} - 1.825 + 0.25 \times 3.825 + 3.825 \times \frac{1}{2}$.

5. $-7.2 \times 0.125 + 0.375 \times 1.1 + 3.6 \times \frac{1}{2} - 3.5 \times 0.375$.

6. $\dfrac{1}{2-\dfrac{1}{3-\dfrac{1}{4-\frac{1}{5}}}}$.

7. $1\frac{1}{2} + 3\frac{1}{4} + 5\frac{1}{8} + 7\frac{1}{16} + 9\frac{1}{32}$.

8. $\frac{1}{1999} + \frac{2}{1999} + \frac{3}{1999} + \cdots + \frac{1998}{1999}$.

9. $(7 + 9 + 11 + \cdots + 101) - (5 + 7 + 9 + \cdots + 99)$.

10. $9 + 99 + 999 + 9999 + 99999 + 999999$.

11. $3^{2000} - 5 \times 3^{1999} + 6 \times 3^{1998}$.

12. $(-1)^{1998} + (-1)^{1999} + (-1)^{2000} + (-1)^{2001}$.

13. $\frac{1}{5 \times 9} + \frac{1}{9 \times 13} + \frac{1}{13 \times 17} + \cdots + \frac{1}{101 \times 105}$.

14. $2002\frac{1}{2} - 2001\frac{1}{3} + 2000\frac{1}{2} - 1999\frac{1}{3} + \cdots + 2\frac{1}{2} - 1\frac{1}{3}$.

15. $\frac{1 \times 2 \times 3 + 2 \times 4 \times 6 + 4 \times 8 \times 12 + 7 \times 14 \times 21}{1 \times 3 \times 5 + 2 \times 6 \times 10 + 4 \times 12 \times 20 + 7 \times 21 \times 35}$.

16. $1 + 2\frac{1}{6} + 3\frac{1}{12} + 4\frac{1}{20} + 5\frac{1}{30} + 6\frac{1}{42} + 7\frac{1}{56}$.

17. $1 + \frac{1}{1+2} + \frac{1}{1+2+3} + \cdots + \frac{1}{1+2+3+\cdots+100}$.

# Chapter 2

# Absolute Value

The absolute value of a positive number is itself; the absolute value of a negative number is its opposite number; the absolute value of zero is zero. This means

$$|a| = \begin{cases} a, & a > 0, \\ 0, & a = 0, \\ -a, & a < 0. \end{cases}$$

The absolute value of a number is the distance between the point corresponding to this number on the axis and the origin. Obviously, the absolute value of any number is a non-negative value, which means $|a| \geqslant 0$.

**Example 1.** What are the integers whose absolute value is equal to 10? What are the integers whose absolute value is less than 10? How many integers have absolute values less than 10? What is their sum?

**Solution** There are two integers, $+10$ and $-10$, whose absolute value is 10. Integers whose absolute value is less than 10 are $0, \pm 1, \pm 2, \ldots, \pm 9$. The total number of them is $2 \times 9 + 1 = 19$. Their sum is

$$0 + 1 + (-1) + 2 + (-2) + \cdots + 9 + (-9) = 0.$$

Remark: Integers are discrete, so the number of integers satisfying the property is finite. If we want to find all rational numbers whose absolute value is less than 10, there will be infinitely many rational numbers.

**Example 2.** Given that $-2 \leqslant a \leqslant 0$. Simplify $|a+2| + |a-2|$.

**Solution**   As $-2 \leqslant a \leqslant 0$, we have $a + 2 \geqslant 0$ and $a - 2 \leqslant 0 - 2 < 0$. Therefore,

$$|a+2| + |a-2|$$
$$= (a+2) - (a-2)$$
$$= 4.$$

**Example 3.** Suppose $x < 0$. Simplify $\frac{||x|-2x|}{|x-3|-|x|}$.

**Solution**   As $x < 0$, so $x - 3 < 0$. Then, we have

$$|x| = -x, |x-3| = -(x-3) = 3 - x,$$
$$|x-3| - |x| = 3 - x - (-x) = 3,$$
$$||x| - 3x| = |-x - 2x| = |-3x| = -3x.$$

Thus, the original formula $= \frac{-3x}{3} = -x$.

Remark:   Given the value or range of a variable, we should first determine the sign of the algebraic formula within the absolute value symbol and then eliminate the absolute value symbol. If there is a multi-layered absolute value symbol, or, in other words, the absolute value symbol contains another absolute value symbol (such as the numerator $||x| - 2x|$ in this example), usually we first eliminate the inner layer and then eliminate the outer layer.

**Example 4.** Suppose $a < 0$ and $x \leqslant \frac{a}{|a|}$. Simplify

$$|x+1| - |x-2|.$$

**Solution**   As $a < 0$, $|a| = -a$, we have $\frac{a}{|a|} = \frac{a}{-a} = -1$.
   Since $x \leqslant \frac{a}{|a|}$, $x \leqslant -1$.
   So, $x + 1 \leqslant 0$ and $x - 2 < 0$.
   Thus,

$$|x+1| - |x-2| = -(x+1) - [-(x-2)]$$
$$= -x - 1 + x - 2 = -3.$$

Remark:   The sign of the number or formula between the absolute value symbols is, sometimes, not given directly by the problem condition. We should first convert the given condition to the condition that we need, just as we have done in this example. Sometimes, the condition is given on the axis, and we should retrieve some useful information from the number line, which is illustrated in the following example.

**Example 5.** As shown in Figure 2.1, the numbers $a$ and $b$ are on the number line. Simplify

$$|a + b| + |b - a| + |b| - |a - |a||.$$

Fig. 2.1

**Solution** As shown in Figure 2.1, we have $a < 0$ and $b > 0$. Furthermore, the distance between $a$ and 0 is greater than the distance between $b$ and 0, thus $a + b < 0$. Then, we have

$$|a + b| + |b - a| + |b| - |a - |a||$$
$$= -(a + b) + (b - a) + b - |a - (-a)|$$
$$= -a - b + b - a + b - (-2a)$$
$$= b.$$

Remark: In this example, from the positions of the numbers $a$ and $b$ on the number line shown in Figure 2.1, we retrieve that $a < 0$, $b > 0$, $a + b < 0$, etc. Then, we can eliminate the absolute value symbol to solve the problem.

**Example 6.** Simplify $\frac{2|x| - 3x}{|2x - |5x||}$.

**Solution** In order to eliminate the absolute value symbol, we have to determine the value of $x$.

Obviously, as the denominator cannot be zero, $x \neq 0$.

If $x > 0$, then we have

$$\frac{2|x| - 3x}{|2x - |5x||} = \frac{2x - 3x}{|2x - 5x|} = \frac{-x}{|-3x|} = \frac{-x}{3x} = -\frac{1}{3}.$$

If $x < 0$, then we have

$$\frac{2|x| - 3x}{|2x - |5x||} = \frac{-2x - 3x}{|2x + 5x|} = \frac{-5x}{|7x|} = \frac{-5x}{-7x} = \frac{5}{7}.$$

Remark: In this example, the sign of the algebraic formula between the absolute value symbols is not given, so we need to categorize different cases. Category discussion is an important idea in mathematics. Problems involving absolute value are good examples of applying this idea. In the following example, the problem contains two absolute value symbols. We categorize the cases by the zero points, which split the number line into several parts.

**Example 7.** Simplify $|x + 5| + |2x - 3|$.

**Analysis**  The key to simplifying this problem is to eliminate the two absolute value symbols. It is easy to eliminate just one absolute value symbol, such as $|x+5|$: we only need to consider the sign of $x+5$ in two cases: $x < -5$ and $x \geqslant -5$. Here, $x = -5$ is the value which makes $x + 5 = 0$. We call it the zero point of $x + 5$. In the same way, for $2x - 3$, it has a zero point at $x = \frac{3}{2}$. In order to eliminate the two absolute value symbols, we mark both zero points $-5$ and $\frac{3}{2}$ on the number line. These two points split the number line into three parts, as shown in Figure 2.2, which are the parts $x < -5$, $-5 \leqslant x < \frac{3}{2}$ and $x \geqslant \frac{3}{2}$. Our discussion is based on these three cases.

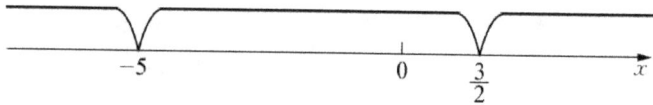

Fig. 2.2

**Solution**   If $x < -5$,

$$\text{the original formula}$$
$$= -(x + 5) - (2x - 3) = -3x - 2.$$

If $-5 \leqslant x < \frac{3}{2}$,

$$\text{the original formula}$$
$$= (x + 5) - (2x - 3) = -x + 8.$$

If $x \geqslant \frac{3}{2}$,

$$\text{the original formula}$$
$$= (x + 5) + (2x - 3) = 3x + 2.$$

We have

$$\text{the original formula} = \begin{cases} -3x - 2, & x < -5, \\ -x + 8, & -5 \leqslant x < \frac{3}{2}, \\ 3x + 2, & x \geqslant \frac{3}{2}. \end{cases}$$

**Example 8.** Simplify $||x - 1| - 2| + |x + 1|$.

**Solution**  We first find the zero points.

From $x - 1 = 0$, we have $x = 1$.

From $|x - 1| - 2 = 0$, which is $|x - 1| = 2$, we have $x - 1 = \pm 2$, $x = -1$, or $x = 3$. From $x + 1 = 0$, we have $x = -1$.

So, there are three zero points: $-1$, $1$, and $3$. They split the number line into four parts, which are

$$x < -1, -1 \leqslant x < 1, 1 \leqslant x < 3, x \geqslant 3.$$

If $x < -1$, then we have

$$\text{the original formula}$$
$$= |-(x - 1) - 2| - (x + 1)$$
$$= |-x - 1| - x - 1$$
$$= -x - 1 - x - 1 = -2x - 2.$$

If $-1 \leqslant x < 1$, then we have

$$\text{the original formula}$$
$$= |-(x - 1) - 2| + x + 1$$
$$= |-x - 1| + x + 1$$
$$= x + 1 + x + 1 = 2x + 2.$$

If $1 \leqslant x < 3$, then we have

$$\text{the original formula}$$
$$= |x - 1 - 2| + x + 1$$
$$= |x - 3| + x + 1$$
$$= 3 - x + x + 1 = 4.$$

If $x \geqslant 3$, then we have

$$\text{the original formula}$$
$$= |x - 1 - 2| + x + 1$$
$$= |x - 3| + x + 1$$
$$= x - 3 + x + 1$$
$$= 2x - 2.$$

Therefore,

$$\text{The original formula} = \begin{cases} -2x - 2, & x < -1, \\ 2x + 2, & -1 \leqslant x < 1, \\ 4, & 1 \leqslant x < 3, \\ 2x - 2, & x \geqslant 3. \end{cases}$$

Remark: In this example, there is a two-layer absolute value symbol. When splitting the number line by zero points, do not forget to consider the zero points of $||x - 1| - 2|$.

**Example 9.** If $|x - 1|$ and $|y + 2|$ are opposite numbers, find $(x + y)^{2002}$.

**Solution**   As $|x - 1|$ and $|y + 2|$ are opposite numbers, we have

$$|x - 1| + |y + 2| = 0.$$

As $|x - 1|$ and $|y + 2|$ are non-negative numbers, we have

$$x - 1 = 0, \quad y + 2 = 0,$$

which is

$$x = 1, \quad y = -2.$$

Then, we have

$$(x + y)^{2002} = (1 - 2)^{2002} = (-1)^{2002} = 1.$$

Remark: If the sum of several non-negative numbers is zero, then each non-negative number is zero. So far, the non-negative numbers we have learned contain absolute value, or they are numbers raised to even powers (mainly square numbers). We have the following.

If $(x - a)^2 + (y - b)^2 = 0$, then we have $x - a = 0$ and $y - b = 0$.
If $|x - a| + (y - b)^2 = 0$, then we have $x - a = 0$ and $y - b = 0$.
If $|x - a| + |y - b| = 0$, then we have $x - a = 0$ and $y - b = 0$.

**Example 10.** Suppose $a$ and $b$ are rational numbers, and $|a + b| = a - b$. Find the value of $ab$.

**Solution**   If $a + b \geqslant 0$, from $|a + b| = a + b = a - b$, we have $b = -b$, then $b = 0$.

If $a + b < 0$, from $|a + b| = -(a + b) = -a - b = a - b$, we have $-a = a$, then $a = 0$.

So, no matter $a + b \geqslant 0$ or $a + b < 0$, there is at least one zero between $a$ and $b$. Thus, $ab = 0$.

## Reading

Socrates' Inspiration

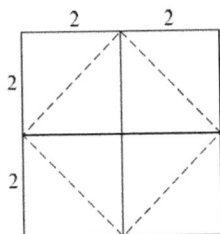

Fig. 2.3

The ancient Greek philosopher Socrates (470–399 BC) liked to use dialogues to inspire students.

On one occasion, he discussed a problem with a boy named Menon: given a square with a side length of 2 feet (with an area of 4 square feet), how to make a square twice the area of the known square.

Menon: Obviously, the side length of this square should be twice that of the known square.

Socrates: Let us draw a square like this (as shown in Figure 2.3).
But there are four squares with a side of 2 feet. Its area is four times that of the known square. Isn't it?

Menon: Yes. But the area of the square we want to get should be twice that of the known square!

Socrates: If we connect the dotted lines in the figure, what figure do these four dotted lines form?

Menon: A square.

Socrates: What is its area?

Menon: I don't know.

Socrates: How many identical triangles are there in this enclosed square?

Menon: There are four of them.

Socrates: How many identical triangles are there in the given square (with a side length of 2 feet) in the upper-left corner?

Menon: There are 2 of them.

Socrates: 4 is how many times 2?

Menon: 2 times 2.

Socrates: From this point of view, how many times is the area of the enclosed square the known square?

Menon: Of course, it is twice the size of the known square. What is the side length of this square with area 2?

---

**Exercises**

1. Determine whether each of the following propositions is true or false:

    (1) If $b < 0$, then $|b| = -b$.          (   )
    (2) If $a$ is a rational number, then $|a|$ must be positive.    (   )
    (3) If $|m| = m$, then $m > 0$.          (   )
    (4) If $a = -b$, then $|a| = |b|$.          (   )
    (5) If $a < b$, then $|a| < |b|$.          (   )
    (6) $a + |a|$ must be positive.          (   )

2. If $-1 < x < 1$, simplify $|x + 1| - |x - 1|$.
3. If $a < 0$, simplify $\frac{2a - |3a|}{||3a| - a|}$.
4. What are the integers whose absolute value is less than 100? How many are there? What is their sum?
5. Simplify $|x - \frac{1}{5}| + |x + \frac{1}{5}|$.
6. Suppose $|a| = 5\frac{2}{3}$, $|b| = 1\frac{1}{3}$, find $a - b$.
7. Suppose $a$ and $b$ are rational numbers. If $a > b$, is $|a| > |b|$ always correct? If it is, please explain the reason; if it is not, please give a counterexample.
8. The rational numbers $a$, $b$, and $c$ are shown in Figure 2.4. Simplify $|a + c| + |b + c| - |a + b|$.

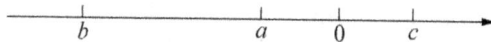

Fig. 2.4

9. If $|a - b| = |a| + |b|$, find the relationship between $a$ and $b$.
10. Suppose $|a + b| + |a - b| = 0$. Simplify

$$|a^{2005} + b^{2005}| + |a^{2005} - b^{2005}|.$$

11. Simplify $|2x - 3| + |3x - 5| - |5x + 1|$.
12. Simplify $||2x - 4| - 6| + |3x - 6|$.
13. Suppose $a$ is a rational number. Find the value of $a + |a|$.

# Chapter 3

# First-Degree Equation with One Variable

Algebraic equations play an important role in secondary school mathematics, and first-degree equations with one variable are the most basic form of algebraic equations.

The standard form of the first-degree equation with one variable is

$$ax = b \ (a \neq 0). \tag{3.1}$$

There is a unique solution to equation (3.1):

$$x = \frac{b}{a}. \tag{3.2}$$

Any first-degree equation with one variable, after simplification, can always be converted into the form $ax = b$.

**Example 1.** Solve the equation $\frac{1}{2}\{x - \frac{1}{3}\left[x - \frac{1}{4}\left(x - \frac{2}{3}\right)\right] - \frac{3}{2}\} = x + \frac{3}{4}$.

**Solution** Multiply both sides of the equation by 2. Then, we have

$$x - \frac{1}{3}\left[x - \frac{1}{4}\left(x - \frac{2}{3}\right)\right] - \frac{3}{2} = 2x + \frac{3}{2}.$$

Move and merge similar terms, and we obtain

$$-\frac{1}{3}\left[x - \frac{1}{4}\left(x - \frac{2}{3}\right)\right] = x + 3.$$

Multiplying both sides by $-3$, we get $x - \frac{1}{4}(x - \frac{2}{3}) = -3x - 9$.
Then, move and merge similar terms, and we have $-\frac{1}{4}(x - \frac{2}{3}) = -4x - 9$.
Multiply both sides by $-4$, and we find $x - \frac{2}{3} = 16x + 36$.

Again, move and merge similar terms to get $-15x = \frac{110}{3}$. By converting the coefficient of the unknown variable to 1, we have $x = -\frac{22}{9}$.

Remark:  In this example, we can also start from inside to outside, eliminating the parentheses, square brackets, and curly brackets in order; however, this method will involve the computation of fractions (just do it as an exercise to improve your ability of fraction computation).

By mental arithmetic (if you are good at it) or by writing down on a piece of paper, you can also compute the coefficient of $x$ and the constant coefficient in the equation. The coefficient of $x$ on the left side is

$$\frac{1}{2}\left\{1 - \frac{1}{3}\left[1 - \frac{1}{4}\right]\right\} = \frac{3}{8}.$$

The constant coefficient is

$$\frac{1}{2}\left\{-\frac{1}{3}\left[\frac{1}{4}\times\frac{2}{3}\right] - \frac{3}{2}\right\} = -\frac{7}{9}.$$

So, we can convert the original equation into

$$-\frac{55}{36} = \frac{5}{8}x,$$

$$x = -\frac{22}{9}.$$

**Example 2.** Solve the equation $\frac{2x+1}{3} - \frac{x-1}{2} = 1$.

**Solution**   Multiply both sides by 6 to get

$$2(2x + 1) - 3(x - 1) = 6,$$

which is $4x - 3x = 6 - 2 - 3$. So, we have $x = 1$.

Remark:  The second-to-last step can be computed by mental arithmetic. It does not require one to write it down on paper.

**Example 3.** When solving the equation $3a - 2x = 15$ (where $x$ is the unknown variable), Bob mistook $-2x$ for $+2x$ and got the solution as $x = 3$. Find the constant $a$ and the solution of the original equation.

**Solution**   According to the problem, the equation Bob solved was actually $3a + 2x = 15$. Since $x = 3$ is a solution to this equation, substitute $x$ for 3 in this equation, and we have

$$3a + 2 \times 3 = 15,$$

and so we have $a = 3$.

Then, the original equation should be $9 - 2x = 15$, and the solution should be $x = -3$.

**Remark:** In this example, we first find $a$ based on the given condition, then find the original equation and its solution. Another method for this example is as follows:

$$3a - 2x = 15,$$
$$3a + 2 \times 3 = 15.$$

Eliminate $a$ to get $\qquad 6 + 2x = 0.$

Thus $\qquad x = -3.$

**Thinking.** The solutions of $3a - 2x = 15$ and $3a + 2x = 15$ are opposite numbers. Why?

**Example 4.** Solve the equation $\frac{x-a}{b} - \frac{x-b}{a} = \frac{b}{a}$ with the unknown variable $x$, where $a \neq 0$, $b \neq 0$ and $a \neq b$.

**Solution** Multiply both sides by $ab$, and we get $a(x-a) - b(x-b) = b^2$.
Then, $(a-b)x + b^2 - a^2 = b^2$.
After moving the terms, we find $(a-b)x = a^2$.
As $a \neq b$, so $a - b \neq 0$, and we have $x = \frac{a^2}{a-b}$.

**Example 5.** Solve the equation $mx - 1 = nx$ with the unknown variable $x$.

**Solution** After moving the terms, we have $(m-n)x = 1$.

(1) If $m - n \neq 0$, which is just $m \neq n$, then the equation has a unique solution $x = \frac{1}{m-n}$.
(2) If $m - n = 0$, which is just $m = n$, as $1 \neq 0$, the equation has no solutions.

**Remark:** When solving a first-degree equation with one variable which includes parameters, we should do a category discussion. The discussion should be integrated and not miss any cases. Next, we present two more examples.

**Example 6.** Solve the equation $4m^2 - x = 2mx + 1$ with the unknown variable $x$.

**Solution** After moving the terms, we get $2mx + x = 4m^2 - 1$, which is simply $(2m+1)x = (2m+1)(2m-1)$.

(1) If $2m + 1 \neq 0$, which is $m \neq -\frac{1}{2}$, then the equation has a unique solution $x = 2m - 1$.
(2) If $2m + 1 = 0$, which is $m = -\frac{1}{2}$, as $(2m+1)(2m-1) = 0$, the equation has infinitely many solutions; furthermore, $x$ can be an arbitrary number.

**Example 7.** Solve the equation $\frac{1}{3}m(x-n) = \frac{1}{4}(x+2m)$ with the unknown variable $x$.

**Solution**  Eliminating the denominator, we find $4mx - 4mn = 3x + 6m$.

After moving terms and merging similar terms, we get $(4m - 3)x = 4mn + 6m$.

(1)  If $4m - 3 \neq 0$, which is $m \neq \frac{3}{4}$, then the original equation has a unique solution: $x = \frac{4mn+6m}{4m-3}$.

(2)  If $4m - 3 = 0$, which is $m = \frac{3}{4}$, then there will be two cases:

If $4mn + 6m = 0$, which is $n = -\frac{6m}{4m} = -\frac{3}{2}$ (since $m = \frac{3}{4} \neq 0$), then the equation has infinitely many solutions, and $x$ can be an arbitrary number.

If $4mn + 6m \neq 0$, which is $n \neq -\frac{6m}{4m} = -\frac{3}{2}$, then the original equation has no solutions.

To conclude,

if $m \neq \frac{3}{4}$, $n$ is an arbitrary number, then the equation has a unique solution: $x = \frac{4mn+6m}{4m-3}$.

If $m = \frac{3}{4}$, $n = -\frac{3}{2}$, then the equation has infinitely many solutions, and $x$ can be an arbitrary number.

If $m = \frac{3}{4}$, $n \neq -\frac{3}{2}$, then the equation has no solutions.

**Example 8.** Given that the equation $2a(x - 1) = (5 - a)x + 3b$ with the unknown variable $x$ has infinitely many solutions, find the values of $a$ and $b$.

**Solution**  After moving the terms, we find $(3a - 5)x = 3b + 2a$.

As the original equation has infinitely many solutions,

$$\begin{cases} 3a - 5 = 0, \\ 3b + 2a = 0. \end{cases}$$

Then, we have $a = \frac{5}{3}$, $b = -\frac{10}{9}$.

Remark:  In this example, the parameters are determined by the features of the solutions.

**Example 9.** Suppose that the equation $ax = b$ with the unknown variable $x$ has two different solutions, $x_1$ and $x_2$. Prove that this equation must have infinitely many solutions.

**Solution**   As $x_1$ and $x_2$ are both solutions to the equation $ax = b$, we have $ax_1 = b$ and $ax_2 = b$, then $ax_1 = ax_2$, which is simply $a(x_1 - x_2) = 0$. As $x_1 \neq x_2$, $a = 0$. Thus,

$$b = ax_1 = 0 \cdot x_1 = 0.$$

As $a = 0$ and $b = 0$, the equation $ax = b$ has infinitely many solutions.

**Another Solution**   If $a \neq 0$, then the equation $ax = b$ has a unique solution: $x = \frac{b}{a}$. But the problem says that the equation $ax = b$ has two different solutions, so we must have $a = 0$. Then, $b = ax_1 = 0 \cdot x_1 = 0$.

As $a = 0$ and $b = 0$, any number can be a solution to $ax = b$.

Remark:   If $a \neq 0$, then the equation $ax = b$ has a unique solution: $x = \frac{b}{a}$.

If $a = 0$ and $b = 0$, then the equation $ax = b$ has infinitely many solutions, and $x$ can be an arbitrary number.

If $a = 0$ and $b \neq 0$, then the equation $ax = b$ has no solutions.

## Reading

### Tian Yuan Technique

In ancient China, the unknown was called Yuan. If there is only one unknown, this unknown is usually called "Tian Yuan."

Write the character "Yuan" next to Tian Yuan, and write the character "Tai" next to the constant term, so that

$$x + 135 = 0$$

can be recorded as in Figure 3.1.

Fig. 3.1

The symbols in the figure are counting chips (small sticks used for calculation). Figure 3.2 represents

$$6x^3 + 2095.4x^2 + 114260.6x - 384133.4 = 0.$$

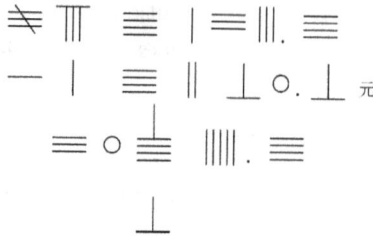

Fig. 3.2

When there are horizontal and vertical counting chips on the same digit, the horizontal one means 5. A slash is drawn on the counting chip to indicate a negative sign.

*Note*: Figure 3.2 is a problem in "Ji Gu Suan Jing" (elaborated in detail by Zhang Dunren).

The Tian Yuan technique was further developed and modified, and its representation method varied over different historical periods.

---

**Exercises**

1. Solve the following equations:

   (1) $3x + 2 = 2x - 5$;

   (2) $3(2x + 1) = 4(x - 3)$;

   (3) $\frac{1}{3}(4 - 3x) = \frac{1}{2}(5x - 6)$;

   (4) $\frac{3}{13}x + \frac{1}{23} = \frac{5}{11}x + \frac{1}{7}$;

   (5) $2x - \frac{2}{3}(x - 2) = \frac{1}{3}[x - \frac{1}{2}(3x + 1)]$;

   (6) $\frac{1}{2}\left\{\frac{1}{2}[\frac{1}{2}(\frac{1}{2}x - 2) - 2] - 2\right\} - 2 = 2$.

2. Solve the following equations with the unknown variable $x$:

   (1) $4mx - 3 = 2x + 6$;

   (2) $4x + b = ax - 8$;

   (3) $9a^2 + 2x = 3ax + 4$;

   (4) $\frac{m}{2}(x + n) = \frac{1}{3}(x + 2)$.

3. Given that the equation $3a(x + 2) = (2b - 1)x + 5$ with the unknown variable $x$ has infinitely many solutions. Find the values of $a$ and $b$.

4. Given that the equation $3x - 3 = 2a(x + 1)$ with the unknown variable $x$ has no solutions. Find the value of $a$.

5. Solve the following equations with the unknown variable $x$:

   (1) $m^2(1 - x) = mx + 1$;

   (2) $(mx - n)(m + n) = 0$.

6. Given that the equation $ax + 3 = 2x - b$ has two different solutions. Find the value of $(a + b)^{2007}$.

7. If the equation $(a + 1)x^2 - 3ax + 2a + 17 = 0$ is a first-degree equation with one variable $x$. Find its solution.

8. Find the natural number $\overline{a_1 a_2 \cdots a_n}$ such that $12 \times \overline{2a_1 a_2 \cdots a_n 1} = 21 \times \overline{1a_1 a_2 \cdots a_n 2}$.

# Chapter 4

# System of First-Degree Equations

A system of first-degree equations is also called a system of linear equations. It is an important tool for solving many practical problems. The basic idea of solving a system of first-degree equations is "eliminating variables." By eliminating one or more variables, the system of first-degree equations is converted into a first-degree equation with one variable for the solution. The commonly used methods of "eliminating variables" include the substitution method and the method of elimination by addition or subtraction.

**Example 1.** Solve the system of equations

$$\begin{cases} x + 3y = 2, & (4.1) \\ 3x - y = -4. & (4.2) \end{cases}$$

**Solution**  By (4.1), we have

$$x = 2 - 3y. \qquad (4.3)$$

Substitute (4.3) in (4.2), and we get

$$3(2 - 3y) - y = -4,$$

$$-10y = -10,$$

$$y = 1.$$

Substitute $y = 1$ in (4.3), and we find $x = -1$.

So, the solution to the original system of equations is

$$\begin{cases} x = -1, \\ y = 1. \end{cases}$$

**Another Solution**   By (4.2), we have $y = 3x + 4$. Then, plug this into (4.1) to eliminate $y$. The rest is left to the readers.

Remark:   We apply the substitution method in this example. It is suitable to use this method when the coefficient of some unknown variable is $\pm 1$.

**Example 2.** Solve the system of equations

$$
\begin{cases}
x + y + z = 3, & (4.4) \\
3x + 2y - z = 1, & (4.5) \\
x - 3y + 2z = 5. & (4.6)
\end{cases}
$$

**Solution**   On (4.4) + (4.5), we have

$$4x + 3y = 4. \tag{4.7}$$

On (4.4) $\times$ 2 $-$ (4.6), we get

$$x + 5y = 1. \tag{4.8}$$

On (4.8) $\times$ 4 $-$ (4.7), we find $17y = 0$, then $y = 0$.
   Substitute $y = 0$ in (4.8) to get $x = 1$.
   Substitute $x = 1$ and $y = 0$ in (4.4) to find $z = 2$.
   So, the solution to the original system of equations is

$$
\begin{cases}
x = 1, \\
y = 0, \\
z = 2.
\end{cases}
$$

Remark:   We apply elimination by addition or subtraction in this example. We first use (4.4) and (4.5), and (4.4) and (4.6) to eliminate $z$ and get a system of equations with variables $x$ and $y$. Then, after solving $x$ and $y$, we substitute them to find $z$. Of course, we can first eliminate either $x$ or $y$. Normally, to solve a system of first-degree equations, we can apply the substitution method, or the method of elimination by addition or subtraction, or even both. These two methods are the most commonly used basic methods. Based on the characteristics of the problems, we can eliminate variables cleverly and simplify the process of solving them.

**Example 3.** Solve the system of equations

$$
\begin{cases}
\dfrac{x}{2} = \dfrac{y}{3} = \dfrac{z}{4}, & (4.9) \\
5x + 2y - 3z = 8. & (4.10)
\end{cases}
$$

**Solution**  Let $\frac{x}{2} = \frac{y}{3} = \frac{z}{4} = k$, which is simply $x = 2k$, $y = 3k$, and $z = 4k$. Substitute them in (4.10), and we have

$$5 \times 2k + 2 \times 3k - 3 \times 4k = 8.$$

Then, we have $k = 2$.

So, we have $x = 4$, $y = 6$, and $z = 8$.

The solution to the original system of equations is

$$\begin{cases} x = 4, \\ y = 6, \\ z = 8. \end{cases}$$

Remark: Setting the ratio as $k$ is a good method when the problem is related to continued proportion.

**Example 4.** Given the system of equations $\begin{cases} 2x - 3y + z = 0, \\ x - 2y + 3z = 0; \end{cases}$ $(xyz \neq 0)$, find $x : y : z$.

**Solution**  Treat the variable $z$ as known, and solve the system of equations with variables $x$ and $y$.

Using the original system of equations, we have to solve the system of equations

$$\begin{cases} 2x - 3y = -z, & (4.11) \\ x - 2y = -3z. & (4.12) \end{cases}$$

On $(4.11) - (4.12) \times 2$, we get

$$y = 5z. \tag{4.13}$$

Substitute (4.13) in (4.12), and we find $x = 7z$.

So, $x : y : z = 7 : 5 : 1$.

Remark: In this example, there are three unknown variables and two equations. Usually, we cannot find the values of $x$, $y$, and $z$. But we can treat one of the unknown variables as known to find $x : y : z$.

**Example 5.** Solve the system of equations

$$\begin{cases} 2x + 3(5x + 7y) = 4, & (4.14) \\ 5x + 7y = 2. & (4.15) \end{cases}$$

**Solution**   Substitute (4.15) into (4.14), and we have

$$2x + 3 \times 2 = 4,$$
$$x = -1. \tag{4.16}$$

Substituting (4.16) in (4.15), we find $y = 1$.

So, the solution to the original system of equations is

$$\begin{cases} x = -1, \\ y = 1. \end{cases}$$

Remark:   Sometimes, according to the characteristics of the problems, we can substitute a whole equation into another to simplify the computation. Of course, not all problems may be solved by substituting some whole equation into another, as done in this example. Sometimes, through careful observation, we can grasp the characteristics of the original system of equations, do some manipulations on them, and then substitute one equation into another.

The general method to solve a system of equations with three variables is to eliminate one unknown variable first to convert it into a system of equations with two variables. But for specific systems of equations, we can use the methods flexibly.

**Example 6.** Solve the system of equations

$$\begin{cases} 5x + 6y - 8z = 12, & (4.17) \\ x + 4y - z = -1, & (4.18) \\ 2x + 3y - 4z = 5. & (4.19) \end{cases}$$

**Solution**   By (4.17), we have

$$x + 2(2x + 3y - 4z) = 12. \tag{4.20}$$

Substitute (4.19) in (4.20), and we get $x = 2$.

Substitute $x = 2$ in (4.18) and (4.19), and we find

$$\begin{cases} 4y - z = -3, & (4.21) \\ 3y - 4z = 1. & (4.22) \end{cases}$$

Substitute (4.21) $\times 4 -$ (4.22) to get $13y = -13$, so $y = -1$.

Substitute $y = -1$ in (4.21) to find $z = -1$.

So, the solution to the original system of equations is

$$\begin{cases} x = 2, \\ y = -1, \\ z = -1. \end{cases}$$

Remark: The substitution of a whole equation is one of the substitution methods, which is similar to eliminating variables. In essence, when solving a system of first-degree equations, the substitution method and the method of elimination by addition or subtraction are both efficient. In this example, we can just compute $(4.17) - (4.19) \times 2$ and get $x = 2$.

**Example 7.** Consider the system of equations with variables $x$ and $y$:

$$\begin{cases} 2x + 3y = 6, & (4.23) \\ ax + 6y = 12. & (4.24) \end{cases}$$

For which values of $a$ does the system of equations have infinitely many solutions? For which values of $a$ does it have one solution?

**Solution**  On $(4.24) - (4.23) \times 2$, we have $(a - 4)x = 0$.

So, if $a - 4 = 0$, which is simply $a = 4$, then $x$ can be any number, and the related $y(= \frac{6-2x}{3})$ has infinitely many choices of values, which means, if $a = 4$, then the original system of equations has infinitely many solutions.

If $a - 4 \neq 0$, which is simply $a \neq 4$, $x = 0 \div (a - 4) = 0$, which means $x$ can only be 0 and the related $y$ should be 2. This means, if $a \neq 4$, the system of equations has one solution: $\begin{cases} x = 0, \\ y = 2. \end{cases}$

Remark: In this example, despite the unknown variables $x$ and $y$, there is another letter $(a)$ in the system of equations. Letters of this type are usually called parameters. We can treat parameters as known, and by using the substitution method and the method of elimination by addition or subtraction, we can convert the system of equations into a first-degree equation with one variable and parameters and then apply the methods introduced in the previous chapter to perform category discussion.

**Example 8.** Solve the system of equations

$$\begin{cases} x + y = 1, & (4.25) \\ y + z = 2, & (4.26) \\ z + x = 3. & (4.27) \end{cases}$$

**Solution**   On (4.25) + (4.26) + (4.27), we have $2(x + y + z) = 6$, which gives

$$x + y + z = 3. \tag{4.28}$$

On $(4.28) - (4.25)$, we have $z = 2$; on $(4.28) - (4.26)$, we find $x = 1$; and on $(4.28) - (4.27)$, we get $y = 0$.

So, the solution to the original system of equations is

$$\begin{cases} x = 1, \\ y = 0, \\ z = 2. \end{cases}$$

**Another Solution**   On $(4.25) + (4.27) - (4.26)$, we have $2x = 2$, then $x = 1$.

Substituting it into (4.25) and (4.27), we get $y = 0$ and $z = 2$.

**Example 9.** Solve the system of equations

$$\begin{cases} \dfrac{4}{x} + \dfrac{6}{y} = 4, & (4.29) \\ \dfrac{1}{3x} + \dfrac{1}{y} = \dfrac{1}{2}. & (4.30) \end{cases}$$

**Solution**   Change the variables by letting $u = \frac{1}{x}$ and $v = \frac{1}{y}$. Then, the system of equations is converted to

$$\begin{cases} 4u + 6v = 4, & (4.31) \\ \dfrac{1}{3}u + v = \dfrac{1}{2}. & (4.32) \end{cases}$$

On $(4.31) - (4.32) \times 6$, we have $2u = 1$, so $u = \frac{1}{2}$.

Substitute $u = \frac{1}{2}$ in (4.32), and we get $v = \frac{1}{3}$.

Then, $\frac{1}{x} = \frac{1}{2}$ and $\frac{1}{y} = \frac{1}{3}$.

So, the solution to the original system of equations is

$$\begin{cases} x = 2, \\ y = 3. \end{cases}$$

Remark:   By changing variables, we convert the original system of equations to a system of first-degree equations.

## Reading

---

### Determinant

$ad - bc$ can be denoted as

$$\begin{vmatrix} a & b \\ c & d \end{vmatrix}, \tag{4.33}$$

and (4.33) is called the (second order) determinant. For example,

$$\begin{vmatrix} 1 & 3 \\ 3 & -1 \end{vmatrix} = 1 \times (-1) - 3 \times 3 = -10. \tag{4.34}$$

Although the determinant is just a notation, it is very convenient to use it to solve equations.

If (4.33) is not 0, then the solution to the system of equations

$$\begin{cases} ax + by = e, & \text{(4.35)} \\ cx + dy = f & \text{(4.36)} \end{cases}$$

is uniquely determined by the method of elimination by addition or subtraction, which is

$$\begin{cases} x = \dfrac{de - bf}{ad - bc}, & \text{(4.37)} \\[3mm] y = \dfrac{af - ce}{ad - bc}. & \text{(4.38)} \end{cases}$$

If written in determinant form, it is easy to memorize:

$$\begin{cases} x = \dfrac{\begin{vmatrix} e & b \\ f & d \end{vmatrix}}{\begin{vmatrix} a & b \\ c & d \end{vmatrix}}, & \text{(4.39)} \\[6mm] y = \dfrac{\begin{vmatrix} a & e \\ c & f \end{vmatrix}}{\begin{vmatrix} a & b \\ c & d \end{vmatrix}}. & \text{(4.40)} \end{cases}$$

Here, the denominators are all "coefficient determinants," that is, (4.33), and the numerator for $x$ is to replace the first column of the coefficient determinant (the coefficients of $x$ in the system of equations) with the constant terms in the system of equations, and the numerator for $y$ is to replace the second column of the coefficient determinant (the coefficients of $y$ in the system of equations) with the constant terms in the system of equations.

For example, in Example 1 in this chapter, as

$$\begin{vmatrix} 2 & 3 \\ -4 & -1 \end{vmatrix} = -2 + 12 = 10,$$

$$\begin{vmatrix} 1 & 2 \\ 3 & -4 \end{vmatrix} = -4 - 6 = -10,$$

so it follows that

$$\begin{cases} x = \frac{10}{-10} = -1, \\ y = \frac{-10}{-10} = 1. \end{cases}$$

The second-order determinants can be used to solve a system of first-degree equations with two unknowns. The third-order determinants can be used to solve a system of first-degree equations with three unknowns.

---

**Exercises**

Solve the following systems of first-degree equations:

1. $\begin{cases} 3x + 5y = 15, \\ 2x - 3y = -4. \end{cases}$

2. $\begin{cases} \frac{x}{3} + \frac{y}{4} = 1, \\ 2x - 3y = 5. \end{cases}$

3. $\begin{cases} x - 2y - 3 = 0, \\ 5x + 4y + 8 = 0. \end{cases}$

4. $\begin{cases} x - 2y + 3z = 0, \\ 3x + 2y + 5z = 12, \\ 2x - 4y - z = -7. \end{cases}$

5. $\begin{cases} 2x - 3y + 4z = 12, \\ x - y + 3z = 4, \\ 4x + y - 3z = -2. \end{cases}$

6. $\begin{cases} x + 2y + 3z = 15, \\ \frac{x+4z}{3} = \frac{y-3z}{4} = 3. \end{cases}$

7. $\begin{cases} 2002x + 2003y = 6007, \\ 2003x + 2002y = 6008. \end{cases}$

8. $\begin{cases} 3(x + 2) + 4y - 10 = 0, \\ 5(x - 2) - 6(y + 2) = 11. \end{cases}$

9. $\begin{cases} x + y + z = 2, \\ y + z + w = 3, \\ z + w + x = 5, \\ w + x + y = 8. \end{cases}$

10. $\begin{cases} 3x + 2y + 2z = 5, \\ 2x + 3y + 2z = 7, \\ 2x + 2y + 3z = 9. \end{cases}$

11. $\begin{cases} 3x - 2y = 3, \\ 7x - 4y = 7. \end{cases}$

12. $\begin{cases} x + y - 3z = 2a, \\ x - 3y + z = 2b, \\ -3x + y + z = 2c. \end{cases}$    ($x$, $y$, and $z$ are the variables.)

13. $\begin{cases} x : y : z = 1 : 3 : 5, \\ x + y + z = 18. \end{cases}$

14. $\begin{cases} \frac{x}{3} = \frac{y}{2} = \frac{z}{5}, \\ 2x + 3y - 4z = 8. \end{cases}$

15. $\begin{cases} \frac{4}{3x} + \frac{3}{2y} = 7, \\ \frac{5}{x} - \frac{6}{y} = 3. \end{cases}$

# Chapter 5

# Application of a System of First-Degree Equations

In this chapter, we discuss some mathematical problems related to the system of first-degree equations.

**Example 1.** Suppose the first-degree equations with two variables $2x + y - 4 = 0$, $x - y + 3 = 0$, and $x + 2y - k = 0$ have a common solution. Find the value of $k$.

**Solution**   Their common solution is the solution to $\begin{cases} 2x + y - 4 = 0, \\ x - y + 3 = 0. \end{cases}$

After solving this system of equations, we have

$$\begin{cases} x = \frac{1}{3}, \\ y = \frac{10}{3}. \end{cases}$$

Substituting it in $x + 2y - k = 0$, we have $\frac{1}{3} + 2 \times \frac{10}{3} - k = 0$, so that $k = 7$.

Remark:   According to the meaning of "common solution," we can begin by solving the system of two equations without $k$, then plug the values in the equation with $k$ and find the value of $k$.

**Example 2.** There is a natural number. If we subtract 63 from it, the result is a perfect square. If we add 26 to it, the result is also a perfect square. Find this natural number.

**Solution**   Suppose this natural number is $x$. The number after adding 26 is $a^2$, and the number after subtracting 63 is $b^2$. Here, $a$ and $b$ are positive integers. Then,

$$\begin{cases} x + 26 = a^2, & (5.1) \\ x - 63 = b^2. & (5.2) \end{cases}$$

On $(5.1) - (5.2)$, we have $89 = a^2 - b^2$, which is simply

$$(a + b)(a - b) = 89. \qquad (5.3)$$

Obviously, $a + b > a - b > 0$. As 89 is a prime number, its divisors are 1 and 89; thus, by (5.3), it must be

$$\begin{cases} a + b = 89, \\ a - b = 1. \end{cases}$$

After solving it, we get $\begin{cases} a = 45, \\ b = 44. \end{cases}$

Thus, $x = 45^2 - 26 = 2025 - 26 = 1999$.

So, the natural number is 1999.

**Remark:**    This problem is an application of the system of first-degree equations in number theory. In this type of problems, usually we get an indeterminate equation or an indeterminate system; however, according to the property of integers (especially representing an integer as a product of prime numbers), we can usually convert it into a system of first-degree equations and then solve it.

**Example 3.** The result of multiplying the sum of two natural numbers by their difference is 84. Find these two natural numbers.

**Solution**    Suppose the larger of these two natural numbers is $x$ and the smaller one is $y$. Then, by the condition, we have $(x + y)(x - y) = 84$.

$84 = 2^2 \times 3 \times 7$. As $(x+y)$ and $(x-y)$ must have the same parity (think about why this is true), together with 84 being even, $(x + y)$ and $(x - y)$ are both even. Since $x + y > x - y$, we have

$$\begin{cases} x + y = 2 \times 3 \times 7, \\ x - y = 2, \end{cases} \quad \text{or} \quad \begin{cases} x + y = 2 \times 7, \\ x - y = 2 \times 3. \end{cases}$$

After solving these two systems of equations, we have $\begin{cases} x = 22, \\ y = 20, \end{cases}$ or $\begin{cases} x = 10, \\ y = 4. \end{cases}$

Thus, the two numbers are 22 and 20, or 10 and 4.

**Example 4.** The value of the algebraic formula $ax^2+bx+c$ is $-2$, 2, and 8 when plugging in $x=0$, $x=1$, and $x=2$, respectively. Find the values of $a$, $b$, and $c$, and also find the value of this algebraic formula when plugging in $x=-1$.

**Solution** By the conditions of the problem, we have

$$\begin{cases} c=-2, \\ a+b+c=2, \\ 4a+2b+c=8. \end{cases}$$

After solving it, we have $a=1$, $b=3$, and $c=-2$.
So, when plugging in $x=-1$, the value of this algebraic formula is

$$1\times(-1)^2+3\times(-1)-2=-4.$$

Remark: This problem is an example of applying the undetermined coefficients method. Using this method (read Chapter 14 for more details), we convert the problem into a system of first-degree equations based on the conditions.

**Example 5.** Consider the system of equations

$$\begin{cases} ax+by=3, \\ 5x-cy=1, \end{cases}$$

After solving it, Bob got the correct answer as $\begin{cases} x=2, \\ y=3. \end{cases}$ Danny mistook $c$ as another number, and he got the solution as $\begin{cases} x=3, \\ y=6. \end{cases}$ Find the values of $a$, $b$, and $c$.

**Solution** Based on the condition, $\begin{cases} x=2, \\ y=3 \end{cases}$ is the solution to the original system of equations. Thus, after plugging in this solution, we have $2a+3b=3$ and $5\times2-3c=1$.

As Danny mistook the value of $c$, $x=3$ and $y=6$ is a solution of $ax+by=3$, giving

$$3a+6b=3.$$

We get a system of equations:

$$\begin{cases} 2a + 3b = 3, \\ 10 - 3c = 1, \\ 3a + 6b = 3. \end{cases}$$

After solving it, we have

$$\begin{cases} a = 3, \\ b = -1, \\ c = 3. \end{cases}$$

So, the values of $a$, $b$, and $c$ are 3, $-1$, and 3, respectively.

**Example 6.** Suppose $|x + y - 4|$ and $(2x - y + 7)^2$ are opposite numbers. Find the values of $x$ and $y$.

**Solution**  Note that $|x+y-4| \geqslant 0$ and $(2x-y+7)^2 \geqslant 0$. As the condition says, they are opposite numbers, they must both be zero, which is simply

$$\begin{cases} x + y - 4 = 0, \\ 2x - y + 7 = 0. \end{cases}$$

After solving it, we have $x = -1$ and $y = 5$.

Remark:   Refer to the remark in Example 9 in Chapter 2.

**Example 7.** Suppose $\begin{cases} x = -1, \\ y = 1, \\ z = 2 \end{cases}$  is a solution to the equation

$$|ax + by + 2| + (ay + cz - 1)^2 + |bz + cx - 3| = 0.$$

Find the values of $a$, $b$, and $c$.

**Solution**  By the condition of this problem, we substitute $x = -1$, $y = 1$, and $z = 2$ in the original equation to get

$$|-a + b + 2| + (a + 2c - 1)^2 + |2b - c - 3| = 0.$$

As $|-a+b+2| \geqslant 0$, $(a+2c-1)^2 \geqslant 0$, and $|2b-c-3| \geqslant 0$, it must be

$$|-a + b + 2| = 0, (a + 2c - 1)^2 = 0, |2b - c - 3| = 0,$$

which is just

$$\begin{cases} -a + b + 2 = 0, \\ a + 2c - 1 = 0, \\ 2b - c - 3 = 0. \end{cases}$$

On solving it, we have $a = 3$, $b = 1$, and $c = -1$.

**Example 8.** Given that $\frac{1}{3}x^{2a+3b}y^{7-a+7b}$ and $-\frac{2}{5}x^{10-b}y^{3a+2b}$ are similar terms, find the values of $a$ and $b$.

**Solution** By the meaning of similar terms, we have

$$\begin{cases} 2a + 3b = 10 - b, \\ 7 - a + 7b = 3a + 2b. \end{cases}$$

After solving it, we have $a = 3$ and $b = 1$.

Remark: This example combines similar terms with the system of first-degree equations.

**Example 9.** Given that $xyz \neq 0$, $x + 2y + z = 0$, and $5x + 4y - 4z = 0$. Find the value of $\frac{x^2 + 6y^2 - 10z^2}{3x^2 - 4yz + 5z^2}$.

**Solution** Treat the variable $z$ as a constant, and let's solve the system of equations with variables $x$ and $y$, which is simply

$$\begin{cases} x + 2y = -z, \\ 5x + 4y = 4z. \end{cases}$$

Then, we find

$$\begin{cases} x = 2z, \\ y = -\frac{3}{2}z. \end{cases}$$

Then, we have the original formula

$$= \frac{(2z)^2 + 6(-\frac{3}{2}z)^2 - 10z^2}{3 \times (2z)^2 - 4 \times (-\frac{3}{2}z)z + 5z^2}$$

$$= \frac{4z^2 + \frac{27}{2}z^2 - 10z^2}{12z^2 + 6z^2 + 5z^2}$$

$$= \frac{15}{46}.$$

Remark: The given algebraic formula contains three variables, but there are only two formulas. Usually, it is not possible to solve to get the values of $x$, $y$, and $z$. However, we can treat $z$ as a constant and obtain a system of equations with variables $x$ and $y$. Then, we can represent $x$ and $y$ by $z$ and substitute them in the algebraic formula to find their values.

**Reading**

---

Ancient Chinese Equations

The eighth volume of the ancient Chinese arithmetic book *The Nine Chapters on the Mathematical Art* is dedicated to "equations." The first problem is as follows:

> Now there are three parcels of upper grade grain, two parcels of middle grade grain, one parcel of lower grade grain, and thirty-nine "dou" in total; two parcels of upper grade grain, three parcels of middle grade grain, one parcel of lower grade grain, and thirty-four "dou" in total; one parcel of upper grade grain, two parcels of middle grade grain, three parcels of lower grade grain, and twenty-six "dou" in total. Here, "dou" is a unit of weight measurement in ancient China. How many "dou" is one parcel of upper, middle, and lower grain, respectively?

The solution to this problem is as follows:

> Suppose that the upper, middle, and lower grains are $x$, $y$, and $z$ "dou" (per parcel), respectively. Then,
> $$\begin{cases} 3x + 2y + z = 39, \\ 2x + 3y + z = 34, \\ x + 2y + 3z = 26. \end{cases}$$
> After solving it, we have $x = 9\frac{1}{4}$, $y = 4\frac{1}{4}$, and $z = 2\frac{3}{4}$.

In ancient times, $x$, $y$, and $z$ were omitted, and the equations were written in the form of a (rectangular) matrix: (The solution is exactly the same as the Gaussian elimination method for solving equations in modern times, and we Chinese had already adopted this amazing method in ancient times.)

$$\begin{pmatrix} 3 & 2 & 1 \\ 2 & 3 & 2 \\ 1 & 1 & 3 \\ 39 & 34 & 26 \end{pmatrix}.$$

The modern way of writing it is

$$\begin{pmatrix} 3 & 2 & 1 & 39 \\ 2 & 3 & 1 & 34 \\ 1 & 2 & 3 & 36 \end{pmatrix}.$$

The columns of the matrix are then transformed (equivalent to elimination) until the matrix becomes

$$\begin{pmatrix} 4 & 0 & 0 \\ 0 & 4 & 0 \\ 0 & 0 & 4 \\ 37 & 17 & 11 \end{pmatrix},$$

meaning $4x = 37$, $4y = 17$, and $4z = 11$, and the solution is $\frac{37}{4}$, $\frac{17}{4}$, and $\frac{11}{4}$.

The process involves operations on different columns, which is called solving (the system of) equations.

---

## Exercises

1. Given the algebraic formula $3ax - b$. If $x = 0$, then its value is 3. If $x = 1$, then its value is 9. Find the values of $a$ and $b$.

2. Given the algebraic formula $ax^2 + 3x - b$. If $x = 1$, then its value is 3. If $x = -2$, then its value is 4. Find its value if $x = 3$.

3. Suppose that $|3x + 2y - 4| + |3y - 2x + 5| = 0$. Find the values of $x$ and $y$.

4. Suppose that $(x - 3y + 6)^2 + |4x - 2y - 3| = 0$. Find the values of $x$ and $y$.

5. Suppose that after multiplying the sum of two natural numbers by their difference, we get the result as 71. Find these two natural numbers.

6. Find all positive integer solutions to the equation $(2x + y)(x - 2y) = 7$.

7. Suppose that $-7x^{2m-2}y^{m-n}$ and $\frac{1}{3}x^{4-m}y^{2n-1}$ are similar terms. Find the values of $m$ and $n$.

8. Given the equation $ax + b = 11$, Bob solved it correctly and got the solution as $x = 3$. Danny mistook $b$ as 6 and got the solution as $x = 5$. Find the values of $a$ and $b$.

9. Given the equation $y = ax^2 + bx + c$ with variables $x$ and $y$. Also, it is given that all of $x = 1$, $y = -2$; $x = 3$, $y = 8$; and $x = -1$, $y = -4$ are solutions to this equation. Find the values of $a$, $b$, and $c$.

10. Suppose that $(3a + 2b - c)^2$ and $|2a + b| + |2b + c|$ are opposite numbers. Find the values of $a$, $b$, and $c$.

11. Find all integer solutions to the equation $(2x - y)(x - 2y) = 5$.

12. Find all positive integer solutions to the equation $xy = x + y$.

13. Find all positive integer solutions to the equation $m^2 - n^2 = 60$.

14. Given the system of equations $\begin{cases} ax + by = 5, \\ cx + dy = -3, \end{cases}$ and its solution is $\begin{cases} x = 2, \\ y = 1. \end{cases}$ Bob mistook $b$ as 6 and got the solution as $\begin{cases} x = 11, \\ y = 1. \end{cases}$ Find the values of $a$, $b$, $c$, and $d$.

# Chapter 6

# Setting Up (Systems of) Equations to Solve Word Problems

Word problems are an important part of middle school mathematics, which help cultivate students' ability to analyze, understand, and solve problems. The main step in solving word problems is to set up equations or a system of equations.

The general steps for solving application problems by setting up (a system of) equations are as follows:

1. Assign the unknown variables according to the problem.
2. Set up some related algebraic formulas.
3. Find the equivalence relation and set up (a system of) equations.
4. Solve (the system of) equations.
5. Check by substitution.
6. Write down the answer.

**Example 1.** It is said that the tombstone of the Greek mathematician Diophantus was engraved with an inscription, translated into simpler words as follows: His childhood took up $\frac{1}{6}$ of his life, the next $\frac{1}{12}$ was adolescence, then $\frac{1}{7}$ of his life passed and he got married. He had a son 5 years later. But the son's fate was not good and only lived to half his age and left in a hurry. He also passed away 4 years later due to excessive grief. How old was Diophantus?

**Solution** Suppose Diophantus had lived for $x$ years. Based on the conditions, we have

$$\frac{1}{6}x + \frac{1}{12}x + \frac{1}{7}x + 5 + \frac{1}{2}x + 4 = x.$$

Eliminating the denominator, we have $14x + 7x + 12x + 420 + 42x + 336 = 84x$.

Move and merge similar terms to get $9x = 756$.

So, $x = 84$.

**Answer** Diophantus had lived for 84 years.

Remark: In this problem, we follow the steps of solving word problems by setting up equations and finding the answer. To find the equivalence relation, we divide the lifespan of Diophantus into several stages, then use the fact that the total amount equals the sum of each component to set up the equation.

**Example 2.** Consider a two-digit number; the sum of the tens digit and the ones digit is 8. Dividing this two-digit number by the difference obtained by subtracting the ones digit from the tens digit, the quotient is 11 and the remainder is 5. Find this two-digit number.

**Solution** Suppose that the tens digit of this two-digit number is $x$ and its ones digit is $(8 - x)$. Then, this two-digit number can be written as $10x + (8 - x)$. According to the given condition, we have

$$10x + (8 - x) = 11 \times [x - (8 - x)] + 5.$$

On solving it, we have $x = 7$, $8 - x = 1$.

**Answer** This two-digit number is 71.

Remark: This is a numerical problem. In this kind of problem, it is usually necessary to express an integer with its digits. Generally, we have

$$\overline{a_1 a_2 a_3 \cdots a_n} = a_n + a_{n-1} \times 10 + a_{n-2} \times 10^2 + \cdots a_2 \times 10^{n-2}$$
$$+ a_1 \times 10^{n-1}$$

(where $\overline{a_1 a_2 a_3 \cdots a_n}$ represents an $n$-digit number and $a_1, a_2, \ldots, a_n$ represent its digits).

**Example 3.** There is a project to build a road. Team A alone takes 10 days, team B alone takes 12 days, and team C alone takes 15 days to complete the work. Now, team A will work for 2.5 days first. Then, team A will stop, and team B will start to work. Finally, there is still a section of road left, and these three teams will work together to complete the task in 2 days. How many days did team B work on the whole project?

**Solution**   Suppose that team B worked for $x$ days on the whole project. Given the conditions, we have

$$\frac{2.5}{10} + \frac{x}{12} + 2 \times \left(\frac{1}{10} + \frac{1}{15}\right) = 1.$$

On solving it, we have $x = 5$.

**Answer**   Team B worked for 5 days on the whole project.

Remark:   This problem is an engineering problem, and the main relations between the quantities used in this kind of problems are:

$$\text{Load} = \text{Efficiency} \times \text{Time};$$
$$\text{Efficiency} = \text{Load} \div \text{Time};$$
$$\text{Time} = \text{Load} \div \text{Efficiency}.$$

If the working load is not given in the problem, we usually regard it as 1.

**Example 4.**   There are three valves: A, B, and C. If we open these three valves at the same time, the pool will be filled in 1 h. If we only open the two valves A and C, the pool will be filled in 1.5 h. If we only open the two valves B and C, the pool will be filled in 2 h. How long will it take to fill the pool, if we only open the two valves A and B?

**Solution**   Suppose if we only open one valve, A, B, or C, the pool will be filled in $x$, $y$, and $z$ h, respectively. Given the conditions, we have

$$\begin{cases} \dfrac{1}{x} + \dfrac{1}{y} + \dfrac{1}{z} = 1, & (6.1) \\[2mm] \dfrac{3}{2}\left(\dfrac{1}{x} + \dfrac{1}{z}\right) = 1, & (6.2) \\[2mm] 2\left(\dfrac{1}{y} + \dfrac{1}{z}\right) = 1. & (6.3) \end{cases}$$

Treat $\frac{1}{x}$, $\frac{1}{y}$, and $\frac{1}{z}$ as the variables. On solving it, we have $\frac{1}{x} = \frac{1}{2}$, $\frac{1}{y} = \frac{1}{3}$, and $\frac{1}{z} = \frac{1}{6}$.

So, we have

$$\frac{1}{x} + \frac{1}{y} = \frac{5}{6}, \qquad (6.4)$$

indicating that if we only open the two valves A and B, the pool will be filled in $1 \div \frac{5}{6} = \frac{6}{5}$(h).

**Answer**   It takes $\frac{6}{5}$ h.

Remark:

1. This problem is actually an engineering problem too.
2. When solving equations, we treat $\frac{1}{x}$, $\frac{1}{y}$ and $\frac{1}{z}$ as the variables, which is actually the idea of changing variables.
3. On $2 \times (6.1) - \frac{2}{3} \times (6.2) - \frac{1}{2} \times (6.3)$, we can get the result of (6.4) directly.

**Example 5.** A certain class of students went to a scenic spot A for a spring outing. The team started at the school and advanced at a speed of 4 km/h. After walking 1 km, the class monitor was sent back to school to pick up a forgotten item. He went back to school at a speed of 5 km/h, and after taking the item, he chased after the team at the same speed and finally caught up with the team at a distance of 1 km from the scenic spot A. Find the distance between the school and the scenic spot A.

**Solution** Suppose that the distance between the school and the scenic spot is $x$ km. Then, from the time when the class monitor left the team to the time when he caught up with the team, he has walked $(x - 1 + 1)$ km and the team has walked $(x - 1 - 1)$ km. As they took the same amount of time to walk, we have

$$(x - 1 + 1) \div 5 = (x - 1 - 1) \div 4.$$

On solving it, we have $x = 10$.

**Answer** The distance between the school and the scenic spot is 10 km.

Remark: This question is an itinerary problem. The main relations used in this type of problems are:

$$\text{Distance} = \text{Speed} \times \text{Time};$$
$$\text{Speed} = \text{Distance} \div \text{Time};$$
$$\text{Time} = \text{Distance} \div \text{Speed}.$$

Itinerary problems include many different types, such as catching up and meeting problems, circular motion problems, running water problems, and uphill and downhill problems. They all have both common and distinct characteristics of the itinerary problems.

**Example 6.** There are only uphill and downhill roads on the trip from A to B, and there are no flat roads. A car travels 20 km/h when going uphill and 35 km/h when going downhill. It takes 9 h for the car to travel from A to B and $7\frac{1}{2}$ h from B to A. How many kilometers is the trip between A and B? How many kilometers are the uphill and the downhill roads from A to B?

**Solution**   Suppose that there are $x$ km uphill road and $y$ km downhill road from A to B. Then, traveling from B to A, there will be $y$ km of uphill road and $x$ km of downhill road. According to the conditions given, we have

$$\begin{cases} \dfrac{x}{20} + \dfrac{y}{35} = 9, \\ \dfrac{x}{35} + \dfrac{y}{20} = 7\dfrac{1}{2}. \end{cases}$$

On solving it, we get

$$\begin{cases} x = 140, \\ y = 70. \end{cases}$$

So, the distance between A and B is $140 + 70 = 210$ (km).

**Answer**   The trip between A and B has 210 km. The uphill road from A to B has 140 km. The downhill road from A to B has 70 km.

**Example 7.**   A farm has two experimental fields, A and B. The area of A is 7 hectares less than half of the total area. The area of B is 32 hectares more than $\frac{1}{3}$ of the total area. How many hectares are there in A and B?

**Solution**   Suppose that the total area of these two experimental fields is $x$ hectares. Then, the area of A is $(\frac{1}{2}x - 7)$ hectares, and the area of B is $(\frac{1}{3}x + 32)$ hectares. Based on the given condition, we have

$$\left(\frac{1}{2}x - 7\right) + \left(\frac{1}{3}x + 32\right) = x.$$

On solving it, we find

$$x = 150.$$

So,

$$\frac{1}{2}x - 7 = 68,$$

$$\frac{1}{3}x + 32 = 82.$$

**Answer**   The area of A is 68 hectares, and the area of B is 82 hectares.

Remark:   In this problem, we apply the method of assigning indirect unknown variables. When solving word problems, there are two methods to assign the unknown variables. The first is the direct method; that is, we define whatever the problem asks as the unknown variables. In all the preceding Examples 1–6, we assign direct unknown variables. The other is the indirect method; that is, sometimes, it is not easy to set up equations with direct unknown variables, but we can define a quantity related to the desired quantity as the unknown variable to find the equation.

This problem can also be solved with a system of equations. The answer is as follows.

**Solution**   Suppose that the area of A is $x$ hectares and the area of B is $y$ hectares. Given the conditions, we have

$$\begin{cases} x = \dfrac{x+y}{2} - 7, \\ y = \dfrac{x+y}{3} + 32. \end{cases}$$

On solving it, we get

$$\begin{cases} x = 68, \\ y = 82. \end{cases}$$

**Answer**   The area of A is 68 hectares, and the area of B is 82 hectares.

**Example 8.** Bookshelves A and B each have a certain number of books. If you take 5 books from shelf B and put them on shelf A, then the number of books on shelf A is 4 times more than that on shelf B. If you take 5 books from shelf A and put them on shelf B, then the number of remaining books on shelf A is 3 times the number on shelf B. How many books are there on shelf A and shelf B?

**Solution**   Suppose that there are $x$ books on shelf A and $y$ books on shelf B. Given the conditions, we have

$$\begin{cases} x + 5 = 5(y - 5), \\ x - 5 = 3(y + 5). \end{cases}$$

On solving it, we get

$$\begin{cases} x = 95, \\ y = 25. \end{cases}$$

**Answer**   There are 95 books on shelf A and 25 books on shelf B.

Remark:

1. If you take 5 books from shelf B to shelf A, then there will be 5 fewer books on shelf B and 5 more books on shelf A. Don't forget the last point.

2. "The number of books on shelf A is 4 times more than that on shelf B" is equivalent to "the number of books on shelf A is 5 times that on shelf B." Generally, the meanings of "$a$ is $k$ times of $b$" and "$a$ is $k$ times more than $b$" are different. The former is equivalent to $a = kb$, and the latter is equivalent to $a = (k+1)b$.

3. This problem can also be solved by assigning one unknown variable and setting up one equation. However, the difficulty of setting up one equation is greater than that of setting up a system of equations. Maybe you should give it a try.

**Example 9.** Bob asks his uncle how old he is. His uncle says, "you would be 4 years old when I was your age. I'll be 40 when you're my age." How old are Bob and his uncle this year?

**Solution One** Suppose that Bob is $x$ years old and his uncle is $y$ years old this year. Clearly, $y > x$. It was $y - x$ years ago when the uncle was Bob's age, and it will be $y - x$ years later when Bob is his uncle's age. Thus,

$$\begin{cases} x - (y - x) = 4, \\ y + (y - x) = 40. \end{cases}$$

On solving it, we get

$$\begin{cases} x = 16, \\ y = 28. \end{cases}$$

**Answer** Bob is 16 years old and his uncle is 28 years old this year.

**Solution Two** Suppose that the difference between their ages is $d$. Then, from the uncle's first statement, we see that Bob is $(4+d)$ years old and his uncle is $(4+2d)$ years old this year. Furthermore, from the uncle's second statement,

$$4 + 3d = 40,$$

so $d = 12$. Then, Bob is $4 + d = 16$ years old, and his uncle is $4 + 2d = 28$ years old.

**Remark:** This problem is related to age. To solve this kind of problem, the key observation is that the difference between two persons' ages is constant at any time. Then, it is easy to set up the equation(s).

**Example 10.** There are four numbers, and the sum of each of the three numbers is 17, 21, 25, and 30, respectively. Find these four numbers.

**Solution One**  Suppose that these four numbers are $x$, $y$, $z$, and $w$. According to the condition, we have

$$\begin{cases} x + y + z = 17, \\ y + z + w = 21, \\ z + w + x = 25, \\ w + x + y = 30. \end{cases}$$

Adding the four equations together and dividing it by 3, we find

$$x + y + z + w = 31.$$

Subtracting each of the four equations, we get

$$\begin{cases} x = 10, \\ y = 6, \\ z = 1, \\ w = 14. \end{cases}$$

**Answer**  These four numbers are 10, 6, 1, and 14.

**Solution Two**  Suppose that the sum of these four numbers is $x$; then, these four numbers will be $(x - 17)$, $(x - 21)$, $(x - 25)$, and $(x - 30)$. Then, we have

$$(x - 17) + (x - 21) + (x - 25) + (x - 30) = x.$$

On solving it, we have $x = 31$.

So, $x - 17 = 14$, $x - 21 = 10$, $x - 25 = 6$, and $x - 30 = 1$.

**Answer**  These four numbers are 14, 10, 6, and 1.

Remark:  In elementary school, our method of solving word problems is mainly arithmetic. After entering middle school, we used the method of setting up (a system of) equations to solve word problems. In some problems where the conditions are obscure or complicated, it is difficult to list the formulas and use arithmetic. By assigning one or more unknown variables, we can make the conditions clearer and translate the mathematical relationship into equations more easily. Furthermore, when it is difficult to solve by introducing just one unknown variable, we can assign several unknown variables to make the equations simpler. Generally, it is easier to solve (a system of) equations than to set (a system of) equations. Examples 8–10 can all be solved by setting only one unknown variable. However, the difficulty of setting up only one equation may be much greater than the difficulty of setting up a system of equations. Maybe you can give it a try.

## Exercises

1. A two-digit number, whose ones digit is 5 greater than the tens digit, is 3 times the sum of its digits. Find this two-digit number.

2. Alice and Bob ride bicycles from $A$ to $B$ at the same time. Alice's speed is 15 km/h, and Bob's speed is 10 km/h. If Alice arrives 10 min before Bob, what is the distance between $A$ and $B$?

3. A project can be completed in 15 days by $A$ alone, 20 days by $B$ alone, and 24 days by $C$ alone. Now, let $A$ and $B$ work together for 5 days, and let $C$ alone complete the rest of the project. How long does $C$ need to complete the project?

4. There are several kilograms of sugar water with a 40% sugar content. After adding 10 kg of water, the sugar concentration becomes 10%. How many kilograms is the original sugar water?

5. Twelve basketball teams have a round-robin match. It is stipulated that each team gets 2 points for a win, 1 point for a loss, and 0 points for abstention. A team participates in all 11 games and gets a total of 17 points. How many games did the team lose and how many games did it win?

6. The distance between $A$ and $B$ is 60 km. A ship travels between $A$ and $B$. It takes 4 h to travel downstream and 5 h to travel upstream. Find the speed of this ship in still water and the speed of running water.

7. A two-digit number, if divided by its ones digit, will have a quotient of 9 and a remainder of 6; if divided by its tens digit, the quotient will be 11 and the remainder will be 1. Find this two-digit number.

8. To manufacture a certain product, if 1 person uses a machine and 3 people rely on handwork, then they can produce 60 pieces per day; if 2 people use machines and 2 people rely on handwork, then they can produce 80 pieces per day. If 3 people use machines and 1 person relies on handwork, how many products can be manufactured per day?

9. Use two solutions with a concentration of 25% and 20% to make 100 g of a solution with a concentration of 22%. How many grams are required for each of the two solutions?

10. $A$ says to $B$: "When I was your age, you would be half of my age; when you were my age, the sum of our ages would be 63 years old." How old are $A$ and $B$ this year?

11. A railway bridge is 1000 m long. A train passes the bridge. It takes 1 min for the train to cross the bridge completely, and the whole train is on the bridge for 40 s. Find the length and speed of the train.

12. The sum of the numbers of people in groups $A$ and $B$ is 28. If group $A$ has 2 more people and group $B$ has 6 more people, then the ratio of the number of people in group $A$ to the number of people in group $B$ will be $2:1$. Find the original numbers of people in groups $A$ and $B$.

13. A person rides a bicycle and starts to move forward at a speed of 15 km/h. When the distance from the destination is 20 km less than the distance traveled, the speed is changed to 10 km/h. In this way, the average speed of the whole journey is 12.5 km/h. How many kilometers is the whole journey?

14. A passenger train and a freight train are running in the same direction on parallel tracks. The length of the passenger train is 200 m, and the length of the freight train is 280 m. The ratio of the speed of the passenger train to the speed of the freight train is $5:3$. When the passenger train catches up with the freight train, the crossing time is just 1 min. Find the speeds of these two trains. If these two trains are running toward each other on parallel tracks, how long will it take for them to cross each other?

15. The first digit of a six-digit number is 1. If this digit is moved to the right of the original ones digit to get a new six-digit number, then the new number is 3 times the original number. Find the original six-digit number.

16. A person walks from A to B. On the first day, he walks several kilometers. If every day from the next day onward, he walks the same distance as the day before, then he can reach B in 10 days. If he walks at the same speed as on the first day, he can reach B in 15 days. If he walks every day at the speed of the last day of the first walking method, how many days will it take to reach B?

# Chapter 7

# (System of) First-Degree Inequalities

An inequality is created by connecting two algebraic formulas with an inequality sign.

The basic properties of inequalities are as follows:

(1) $a - b > 0 \Leftrightarrow a > b$; $a - b < 0 \Leftrightarrow a < b$.

(2) $a > b \Leftrightarrow b < a$; $a < b \Leftrightarrow b > a$.

(3) $a > b, b > c \Rightarrow a > c$.

(4) $a > b \Leftrightarrow a + c > b + c$.

(5) $a > b, c > 0 \Rightarrow ac > bc$;
$\quad a > b, c < 0 \Rightarrow ac < bc$.

(6) $a > b > 0, c > d > 0 \Rightarrow ac > bd$.

Although these properties are simple, they are the basis for solving inequalities. So, please get familiar with them as soon as possible. Among them, property (5) deserves special attention. If there is only one unknown variable in the inequality and the highest degree of the unknown variable is 1, such an inequality is called a first-degree inequality with one variable. Its general form is $ax > b$ (or $ax < b$, $ax \geqslant b$, $ax \leqslant b$), where $a \neq 0$. Any first-degree inequality with one variable can be converted into a general form by using the above basic properties.

A value of $x$ that makes the inequality hold is called a solution to the inequality. The set of all solutions is called the solution set of the inequality. This chapter mainly discusses how to solve (a system of) first-degree inequalities.

**Example 1.** Solve the inequality $\frac{x-2}{3} - x \geqslant \frac{x-1}{2}$.

**Solution**  Multiply by 6 on both sides to get

$$2(x-2) - 6x \geqslant 3(x-1),$$

which is simply

$$-4x - 4 \geqslant 3x - 3.$$

Move and merge items, and we have

$$-7x \geqslant 1.$$

Dividing both sides by $-7$ (pay attention to the change in the direction of the inequality sign), we have $x \leqslant -\frac{1}{7}$.

So, the solution to the original inequality is $x \leqslant -\frac{1}{7}$.

Remark:  The method of solving a first-degree inequality is similar to the method of solving a first-degree equation; just pay attention to the direction of the inequality sign. When both sides of the inequality are multiplied or divided by a positive number at the same time, the direction of the inequality sign remains unchanged. But when both sides of the inequality are multiplied or divided by a negative number, the direction of the inequality sign changes.

After converting a first-degree inequality to the general form $ax < b$, using the property of inequality, we have that

(1) if $a > 0$, then the solution is $x < \frac{b}{a}$;
(2) if $a < 0$, then the solution is $x > \frac{b}{a}$;
(3) if $a = 0$ and $b > 0$, then the solution includes all numbers; if $a = 0$ and $b \leqslant 0$, then the inequality has no solutions.

There are similar conclusions for the inequality $ax \leqslant b$, $ax > b$, and $ax \geqslant b$. You may find their properties by yourself.

**Example 2.** Solve the inequality with variable $x$:

$$2mx + 3 < 3x + n.$$

**Solution**  From the original inequality, we get

$$(2m - 3)x < n - 3.$$

(1) If $2m - 3 > 0$, which is $m > \frac{3}{2}$, then the solution is $x < \frac{n-3}{2m-3}$.
(2) If $2m - 3 < 0$, which is $m < \frac{3}{2}$, then the solution is $x > \frac{n-3}{2m-3}$.

(3) If $2m - 3 = 0$, which is $m = \frac{3}{2}$, then there are two cases:

If $n - 3 > 0$, which is $n > 3$, the solution contains all numbers.
If $n - 3 \leqslant 0$, which is $n \leqslant 3$, the original inequality has no solutions.

Remark: Like equations, letters in inequalities that are not unknown variables are called parameters. When solving inequalities with parameters, the parameters' conditions should also be discussed.

**Example 3.** For which value(s) of $k$, the equation with variable $x$,

$$5(x - k) = 3x - k + 2,$$

has

(1) a positive solution;
(2) a negative solution;
(3) a solution not greater than 1.

**Solution** From the original equation, we have $2x = 4k + 2$, $x = 2k + 1$.

(1) The equation has a positive solution, which is $2k + 1 > 0$. Thus, when $k > -\frac{1}{2}$, the equation has a positive solution.
(2) The equation has a negative solution, which is $2k + 1 < 0$. Thus, when $k < -\frac{1}{2}$, the equation has a negative solution.
(3) The equation has a solution not greater than 1, which is $2k + 1 \leqslant 1$. Thus, when $k \leqslant 0$, the equation has a solution not greater than 1.

**Example 4.** Solve the system of inequalities

$$\begin{cases} 3x + 2 > 2x - 5, \\ x + 4 > \frac{x}{2} + 9. \end{cases}$$

**Solution** After solving the two inequalities separately, we have $x > -7$ and $x > 10$. To make these two formulas both hold, it must be $x > 10$. So, the solution to the original system of inequalities is $x > 10$.

Remark: A system of first-degree inequalities with one variable consists of several first-degree inequalities, and its solution set is the common part of the solution sets of all the inequalities.

If the system consists of two inequalities, we first solve the two inequalities separately. Now, there are four cases, as follows (where we suppose that $a < b$):

(1) The solution to $\begin{cases} x > a, \\ x > b \end{cases}$ is $x > b$. (See Figure 7.1.)

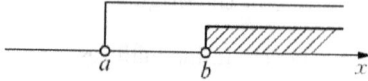

Fig. 7.1

(2) The solution to $\begin{cases} x < a, \\ x < b \end{cases}$ is $x < a$. (See Figure 7.2.)

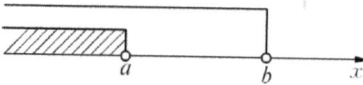

Fig. 7.2

(3) The solution to $\begin{cases} x > a, \\ x < b \end{cases}$ is $a < x < b$. (See Figure 7.3.)

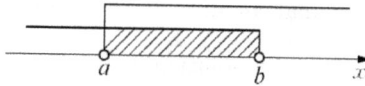

Fig. 7.3

(4) $\begin{cases} x < a, \\ x > b \end{cases}$ has no solution. (See Figure 7.4.)

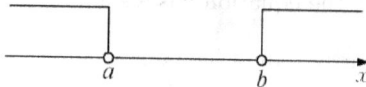

Fig. 7.4

If the system of inequalities contains more than two inequalities, we can first solve each inequality and obtain

$$\begin{cases} x > a_1, \\ x > a_2, \\ \cdots \\ x > a_n, \\ x < b_1, \\ x < b_2, \\ \cdots \\ x < b_m. \end{cases}$$

We must determine its upper bound $b$, which is the smallest among $b_1$, $b_2, \ldots, b_m$; and also determine its lower bound $a$, which is the greatest among $a_1, a_2, \ldots, a_n$.

If $a < b$, then the solution to the original system of inequalities is $a < x < b$.

If $a \geqslant b$, then the original system of inequalities has no solutions.

**Example 5.** Solve the system of inequalities

$$\begin{cases} 9x - 3 > x + 8, \\ -3x + 40 > x + 12, \\ \frac{x}{2} - 1 < \frac{x}{3}, \\ \frac{2x-3}{3} > 1. \end{cases}$$

**Solution** From the original system of inequalities, we have

$$\begin{cases} x > \frac{11}{8}, \\ x < 7, \\ x < 6, \\ x > 3. \end{cases}$$

Determine the upper bound: as $x < 7$ and $x < 6$, we have $x < 6$. Determine the lower bound: ay $x > \frac{11}{8}$ and $x > 3$, we have $x > 3$.

So, the solution to the original system of inequalities is $3 < x < 6$.

**Example 6.** Solve the following inequalities:

(1) $4x - 2 + \frac{1}{x-5} > \frac{1}{x-5} + 3x + 2$;

(2) $\frac{7x-6}{2x+3} > 2$.

**Solution**

(1) Eliminate $\frac{1}{x-5}$ on both sides, and we get $4x - 2 > 3x + 2$, $x > 4$.
    But note that in the original inequality, $x - 5 \neq 0$, which is $x \neq 5$. So, we should delete $x = 5$ from $x > 4$. The solution to the original inequality is $x > 4$ and $x \neq 5$.

(2) Multiply by $2x + 3$ on both sides to eliminate the denominator.
    If $2x + 3 > 0$, which is $x > -\frac{3}{2}$, after eliminating the denominator, we have $7x - 6 > 4x + 6$, so $x > 4$. Combined with $x > -\frac{3}{2}$, we have $x > 4$.
    If $2x + 3 < 0$, which is $x < -\frac{3}{2}$, after eliminating the denominator, we have $7x - 6 < 4x + 6$. So, $x < 4$. Combined with $x < -\frac{3}{2}$, we have $x < -\frac{3}{2}$. So, the solution to the original inequality is $x > 4$ or $x < -\frac{3}{2}$.

**Remark:** The above two inequalities are not first-degree inequalities, but they can be converted into (a system of) first-degree inequalities. Although we can eliminate $\frac{1}{x-5}$ on both sides of the first inequality, we must pay attention to $x - 5 \neq 0$. For the second inequality, when multiplied by the denominator, there are two cases to discuss.

**Example 7.** The solution to the system of inequalities with variable $x$,

$$\begin{cases} \dfrac{x+4}{3} - 1 > \dfrac{x}{2}, & (7.1) \\ x + k < 0, & (7.2) \end{cases}$$

is $x < 2$. Find the range of $k$.

**Solution**  We first solve the original system of inequalities.

By (7.1), we have $x < 2$. By (7.2), we have $x < -k$.

As the solution to the system of inequalities is $x < 2$, it must be that $2 \leqslant -k$, and thus the range of $k$ is $k \leqslant -2$.

**Example 8.** Solve the system of inequalities with variable $x$:

$$\begin{cases} ax - 4 < 8 - 3ax, \\ (a + 2)x - 2 > 2(1 - a)x + 4. \end{cases}$$

**Solution**  The original system of inequalities can be converted into

$$\begin{cases} ax < 3, & (7.3) \\ ax > 2. & (7.4) \end{cases}$$

If $a = 0$, then by (7.4), the system of inequalities has no solutions.
If $a > 0$, then by (7.3) and (7.4), we have

$$\frac{2}{a} < x < \frac{3}{a}.$$

If $a < 0$, then by (7.3) and (7.4), we have

$$\frac{3}{a} < x < \frac{2}{a}.$$

**Exercises**

1. Solve the following inequalities:
   (1) $2(x - 1) - 3x > 4(x + 1) + 5$;
   (2) $\frac{x+1}{3} - \frac{x-3}{2} > 5$;

(3) $\frac{1}{3}x - 3 + \frac{2x-6}{4} < -3x$;

(4) $3x + \frac{1}{x-2} + 2 > x + 4 + \frac{1}{x-2}$;

(5) $\frac{2x-1}{3} - 1 \geqslant \frac{3x+2}{6} + \frac{x}{2}$;

(6) $5 - \frac{x}{2} \geqslant 3\frac{1}{2} - (\frac{4x+1}{8} - \frac{x+2}{4})$.

2. Solve the following systems of inequalities:

(1) $\begin{cases} 2x - 2 > \frac{1}{2}(x - 7) + 1, \\ \frac{2x-5}{3} + 3 < \frac{x}{5} + 2; \end{cases}$

(2) $\begin{cases} 4x + \frac{2}{3} < \frac{x-8}{5} + 2, \\ 2 - \frac{x}{3} > 3 - \frac{x}{2}; \end{cases}$

(3) $\begin{cases} 1 < \frac{2x-5}{3} < 3, \\ (x - \frac{2}{3}) + 5 \geqslant 2x - \frac{3}{2}; \end{cases}$

(4) $\begin{cases} x - 1 > -3, \\ \frac{x}{2} - 1 < \frac{x}{3}, \\ 3 < 2(x - 1) < 10; \end{cases}$

(5) $\begin{cases} 2x + 3 < 9 - x, \\ 6x - 1 < 5, \\ 2 - x \leqslant 3x + 7. \end{cases}$

3. For which values of $k$, does the equation $5(x + 3k) - 2 = 3x - 4k$ with variable $x$ have

(1) a positive solution;

(2) a negative solution?

4. Solve the system of inequalities with variable $x$:

$$\begin{cases} a(x - 2) > x - 3, \\ 9(a + 1)x > 9ax + 8. \end{cases}$$

5. For which values of $k$, does the equation $3k(x + 2) + 2 = 5x + 8$ with variable $x$ have a solution not greater than 3?

6. Prove the inequality that $\frac{1}{2^2} + \frac{1}{3^2} + \cdots + \frac{1}{n^2} < \frac{n-1}{n}$, for any integer $n \geqslant 2$.

# Chapter 8

# Multiplication and Division of Polynomials with Integer Coefficients

Multiplication and division of polynomials with integer coefficients are two basic operations in polynomials with integer coefficients. Similar to multiplication of numbers, multiplication of polynomials with integer coefficients also satisfies the commutative, associative, and distributive properties. The division of polynomials with integer coefficients can be regarded as the inverse operation of multiplication and is similar to the division of numbers.

Based on the properties of multiplication, people have summarized some useful multiplication formulas.

**Example 1.** Compute $(-x^3 + 2x^2 - 5)(2x^2 - 3x + 1)$.

**Solution**

The original formula
$$= -2x^5 + 3x^4 - x^3 + 4x^4 - 6x^3 + 2x^2 - 10x^2 + 15x - 5$$
$$= -2x^5 + 7x^4 - 7x^3 - 8x^2 + 15x - 5.$$

Remark: When multiplying two polynomials, choose every pair of terms from the polynomials and multiply, being careful not to repeat or omit any pair (for example, you can first multiply $2x^2$, $-3x$, and $1$ by $-x^3$, then multiply $2x^2$, $-3x$, and $1$ by $2x^2$ and $-5$, respectively), and then merge the similar terms to get the final result. If you are proficient, you can do mental calculations, arrange the terms in the descending order of the powers of $x$, and write out the terms of $x$ one by one, then merge the similar terms. In addition, comparing the original formula and the result, you will find the highest degree term and the constant term are exactly the product of the highest degree terms of two factors and the product of two constant terms, respectively (which are $-2x^5 = -x^3 \cdot 2x^2$ and $-5 = -5 \times 1$). Comparing the same degree terms of two algebraic formulas, especially the highest degree terms and the constant terms, is frequently used in mathematics competitions. We show it in the following example.

**Example 2.** Given that $(ax^3 - x + 6)(3x^2 + 5x + b) = 6x^5 + 10x^4 - 7x^3 + 13x^2 + 32x - 12$, find the values of $a$ and $b$.

**Solution**   By the condition, we compare the highest degree terms and the constant terms, and we have

$$a \cdot 3 = 6, \quad 6 \cdot b = -12.$$

So, $a = 2$ and $b = -2$.

Remark:   It is not necessary to compute $(ax^3 - x + 6)(3x^2 + 5x + b)$.

**Example 3.** Compute $(3x^2 + 2)(5x^4 + 2x^2 + 3) - (5x^4 + x^2 + 3)(3x^2 + 3)$.

**Solution**

The original formula
$$= (3x^2 + 2)(5x^4 + 2x^2 + 3) - (5x^4 + x^2 + 3)(3x^2 + 2 + 1)$$
$$= (3x^2 + 2)(5x^4 + 2x^2 + 3) - (5x^4 + x^2 + 3)(3x^2 + 2) - (5x^4 + x^2 + 3)$$
$$= (3x^2 + 2)[(5x^4 + 2x^2 + 3) - (5x^4 + x^2 + 3)] - (5x^4 + x^2 + 3)$$
$$= x^2(3x^2 + 2) - (5x^4 + x^2 + 3)$$
$$= 3x^4 + 2x^2 - 5x^4 - x^2 - 3$$
$$= -2x^4 + x^2 - 3.$$

Remark:   We can use the multiplication properties (of polynomials with integer coefficients) to simplify the computation.

**Example 4.** Prove that the sum of squares of any five consecutive integers is not a square number.

**Solution**   Suppose that these five consecutive integers are $n-2$, $n-1$, $n$, $n+1$, and $n+2$ (where $n$ is an integer). Then,

$$(n-2)^2 + (n-1)^2 + n^2 + (n+1)^2 + (n+2)^2$$
$$= n^2 - 4n + 4 + n^2 - 2n + 1 + n^2 + n^2 + 2n + 1 + n^2 + 4n + 4$$
$$= 5n^2 + 10$$
$$= 5(n^2 + 2).$$

The ones digit of a square number can only be 0, 1, 4, 5, 6, or 9, and so the ones digit of $n^2 + 2$ can only be 2, 3, 6, 7, 8, or 1. So, $n^2 + 2$ cannot be divided by 5. Then, $5(n^2 + 2)$ can be divided by 5 but cannot be divided by $5^2$, so it is not a square number.

**Example 5.** Compute $(3x^4 - 5x^3 + x^2 + 2) \div (x^2 + 3)$.

**Solution**

$$
\begin{array}{r}
3x^2 - 5x - 8 \\
x^2+3 \overline{)3x^4 - 5x^3 + x^2 + 0x + 2} \\
\underline{3x^4 \qquad\quad + 9x^2} \\
-5x^3 - 8x^2 + 0x \\
\underline{-5x^3 \qquad\quad - 15x} \\
-8x^2 + 15x + 2 \\
\underline{-8x^2 \qquad\quad - 24} \\
15x + 26.
\end{array}
$$

The quotient is $3x^2 - 5x - 8$, and the remainder is $15x + 26$.

Remark:   The division of polynomials can be computed by long division of polynomials. During computation, pay attention to the descending power arrangement, fill 0 (or blank space) for the missing terms, align terms with the same degree, etc.

For long division of polynomials, we have a quotient and a remainder, that is, Dividend = Divisor × Quotient + Remainder, where the highest degree of the remainder is lower than the highest degree of the divisor. When the remainder is 0, we also call it divisible, and it is represented by "Divisor|Dividend." For example, $x-1|x^3-1$. When the remainder is not 0, we say that the dividend is not divisible; when the remainder is a constant, we also call it the remainder number. Obviously, when the divisor is a first-degree polynomial, the remainder must be a constant.

**Example 6.** Compute $(2x + 1) \div (3x - 2) \times (6x - 4) \div (4x + 2)$.

**Solution**

The original formula
$$= [(2x + 1) \div (4x + 2)] \times [(6x - 4) \div (3x - 2)]$$
$$= (2x + 1) \div [2(2x + 1)] \times [2(3x - 2) \div (3x - 2)]$$
$$= \frac{1}{2} \times 2$$
$$= 1.$$

**Remark:** In complicated multiplications and divisions, sometimes a clever use of the associative property can simplify the calculations.

According to the properties of polynomial multiplication, people have summarized some multiplication formulas as follows. We can use these formulas to make the calculations faster and more accurate:

(1) $(a \pm b)^2 = a^2 \pm 2ab + b^2$.
(2) $(a \pm b)^3 = a^3 \pm 3a^2b + 3ab^2 \pm b^3$.
(3) $(a + b)(a - b) = a^2 - b^2$.
(4) $(a \pm b)(a^2 \mp ab + b^2) = a^3 \pm b^3$.
(5) $(a + b + c)^2 = a^2 + b^2 + c^2 + 2ab + 2bc + 2ca$.
(6) $(a - b)(a^{n-1} + a^{n-2}b + a^{n-3}b^2 + \cdots + ab^{n-2} + b^{n-1}) = a^n - b^n$.

Once you become familiar with the above formulas and use them flexibly (in either direction), you can solve some problems skillfully.

**Example 7.** Compute $(a - b)(a + b)(a^2 + b^2)(a^4 + b^4)$.

**Solution**

The original formula
$$= (a^2 - b^2)(a^2 + b^2)(a^4 + b^4)$$
$$= (a^4 - b^4)(a^4 + b^4)$$
$$= a^8 - b^8.$$

Remark: This problem can be extended to the following:

$$(a - b)(a + b)(a^2 + b^2)(a^4 + b^4) \cdots (a^{2^n} + b^{2^n})$$
$$= (a^2 - b^2)(a^2 + b^2)(a^4 + b^4) \cdots (a^{2^n} + b^{2^n})$$

$$\cdots$$

$$= (a^{2^n} - b^{2^n})(a^{2^n} + b^{2^n})$$
$$= a^{2^{n+1}} - b^{2^{n+1}}.$$

**Example 8.** Find the remainder number of $x^{285} - x^{83} + x^{71} + x^9 - x^3 + x$ when divided by $x - 1$.

**Solution**

$$x^{285} - x^{83} + x^{71} + x^9 - x^3 + x$$
$$= (x^{285} - 1) - (x^{83} - 1) + (x^{71} - 1) + (x^9 - 1)$$
$$- (x^3 - 1) + (x - 1) + 2.$$

By the multiplication formula (6), $x^{285} - 1$, $x^{83} - 1$, $x^{71} - 1$, $x^9 - 1$, $x^3 - 1$, and $x - 1$ are all divisible by $x - 1$. So, the remainder number after the division by $x - 1$ is 2.

Remark: The degree of $x^{285}$ is too high, and it is obviously difficult to obtain the quotient by the long division method. Add $-1$ after $x^{285}$, and use formula (6), $x - 1$ divides $x^{285} - 1^{285} = x^{285} - 1$. Similarly, add $-1$ after every power of $x$. Finally, to keep the formula unchanged, we must add 2 at the end. In this way, the binomial in each bracket is divisible by $x - 1$, and so the remainder number after the division by $x - 1$ is 2.

We used formula (6) in the opposite direction in this problem. Note that each formula can be used either forward or backward. Use them wisely to solve the problems.

## Reading

---

### Separating Coefficients Method

During multiplication and division of polynomials of integer coefficients, if there is only one variable (letter), long multiplication or long division is commonly used, where the letters are usually omitted (but attention should be paid to arranging the terms in descending or ascending powers and leaving the positions of the missing terms). This is called the "separating coefficients method."

For example, Example 1 of this chapter can be written as shown in Figure 8.1.

$$
\begin{array}{r}
-1+\ 2+\ 0-5 \\
\times \qquad 2-\ 3+1 \\
\hline
-1+\ 2+\ 0-5 \\
3-6+\ 0+15 \\
-2+4+0-10 \\
\hline
-2+7-7-\ 8+15-5
\end{array}
$$

Fig. 8.1

Also, Example 5 of this chapter can be written as shown in Figure 8.2.

$$
\begin{array}{r}
3-\ 5-\ 8 \\
1+0+3\ )\ 3-5+1+\ 0+\ 2 \\
3+0+9 \\
\hline
-5-8+\ 0 \\
-5+0-15 \\
\hline
-8+15+\ 2 \\
-8+\ 0-24 \\
\hline
15+26
\end{array}
$$

Fig. 8.2

## Exercises

1. Compute the following formulas:

  (1) $(3x^2 + 2x - 3)(2x - 1)$;
  (2) $(4x^4 - 6x^2 + 2)(5x^3 - 2x^2 + x - 1)$;
  (3) $(a + b)^2 - (a - b)^2$;
  (4) $(a + b)^3 - 3ab(a + b)$;
  (5) $(a + b + c)(a^2 + b^2 + c^2 - ab - bc - ca)$;
  (6) $(3x^3 - 4x^2 + 5x - 1) \div (x^2 + 3x - 1)$;
  (7) $(5x^3 - 7x + 1) \div (2x + 1)$;
  (8) $(x^3 + 1) \div (x + 1)$;

(9) $(a^2 - b^2) \div (a^2 + 2ab + b^2) \times (a^3 + b^3)$;

(10) $(7x^2 + 3x) \div (2x + 1) \times (6x + 3) \div (7x + 3)$.

2. Use the multiplication formulas to compute the following formulas:

(1) $2000^2 - 1999 \times 2001$;

(2) $(2 + 1)(2^2 + 1)(2^4 + 1) \cdots (2^{2^n} + 1)$.

3. Find the remainder of $x^{128} + x^{110} - x^{32} + x^8 + x^2 - x$ when divided by $x - 1$.

4. Find the remainder of $x^{111} - x^{31} + x^{13} + x^9 - x^3$ when divided by $x + 1$.

5. Prove that the difference between the squares of two adjacent odd integers is a multiple of 8 and is equal to twice their sum.

6. Prove that if $a + b + c = 0$, then $a^3 + b^3 + c^3 = 3abc$.

7. Prove that

$$(x + y - 2z)^3 + (y + z - 2x)^3 + (z + x - 2y)^3$$

$$= 3(x + y - 2z)(y + z - 2x)(z + x - 2y).$$

# Chapter 9

# Line Segments

Line segments are a basic figure of geometry. Various geometric figures can be formed by line segments, such as triangles and quadrilaterals. Therefore, it is very important to observe the line segments in the figure and find the relationship between them and other shapes.

This chapter introduces some knowledge related to line segments, such as that the line segment between two points is the shortest and how to calculate the lengths of some given line segments.

When comparing the lengths of line segments, a very useful fact is that the line segment between two points is the shortest. In other words, among all the lines connecting two points, the length of the line segment is the shortest. From this fact, we get an important corollary: the sum of any two side lengths in a triangle is greater than the third side length. That is, as shown in Figure 9.1, we have

$$AB + AC > BC,$$
$$AB + BC > AC,$$
$$AC + BC > AB.$$

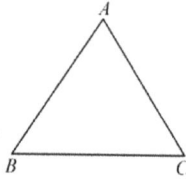

Fig. 9.1

This corollary is very useful, which is shown in the following example.

**Example 1.** Given that $O$ is an arbitrary point in $\triangle ABC$ (shown in Figure 9.2). Prove that

$$\frac{1}{2}(AB + AC + BC) < OA + OB + OC.$$

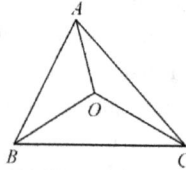

Fig. 9.2

**Proof**  In $\triangle OAB$, $OA + OB > AB$.
In $\triangle OAC$, $OA + OC > AC$.
In $\triangle OBC$, $OB + OC > BC$.
Adding these three formulas, we have

$$2(OA + OB + OC) > AB + AC + BC,$$

which is simply

$$\frac{1}{2}(AB + AC + BC) < OA + OB + OC.$$

**Example 2.** In $\triangle ABC$, $D$ is the midpoint of $BC$. Prove that

$$AB + AC > 2AD. \tag{9.1}$$

**Analysis** We try to construct a triangle, the lengths of whose three sides are $AB$, $AC$, and $2AD$.

**Solution** As shown in Figure 9.3, put $\triangle ABD$ under the line $DC$ to make

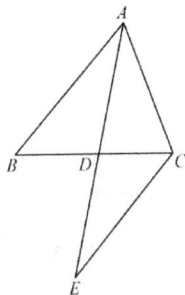

Fig. 9.3

the points $D$ and $D$ coincide and the points $B$ and $C$ coincide (as $BD = DC$, this is feasible).

As $\angle EDC = \angle ADB$ and $\angle EDC + \angle ADC = \angle ADB + \angle ADC = 180°$, the points $A$, $D$, and $E$ are collinear: $AE = AD + DE = 2AD$.

In $\triangle AEC$, $AC + EC > AE$.

As $EC = AB$, the above inequality is the same as (9.1).

**Example 3.** As shown in Figure 9.4, points A, B, C, and D in the figure are four production workshops of a factory. Now, a warehouse needs to be built somewhere such that the sum of the distances from the warehouse to the four production workshops A, B, C, and D is minimized. Where should the warehouse be built?

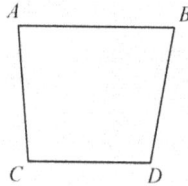

Fig. 9.4

**Solution**   As shown in Figure 9.5, the warehouse should be built at the intersection point $P$ of the diagonals $AD$ and $BC$. We prove that $PA + PB + PC + PD$ is the smallest.

Suppose that point $P'$ is a different point of $P$. We need to prove

$$PA + PB + PC + PD < P'A + P'B + P'C + P'D.$$

In $\triangle P'BC$, $BC \leqslant P'C + P'B$, which is $PB + PC \leqslant P'B + P'C$. Equality holds if and only if point $P'$ lies on the line segment $BC$.

In $\triangle P'AD$, $AD \leqslant P'A + P'D$, which is $PA + PD \leqslant P'A + P'D$. Equality holds if and only if point $P'$ lies on the line segment $AD$.

Adding the two inequalities, we have

$$PA + PB + PC + PD \leqslant P'A + P'B + P'C + P'D.$$

When the equality holds, we get that point $P'$ lies both on line segments $AD$ and $BC$. Then, $P = P'$. Since point $P'$ is a different point of $P$, we get

$$PA + PB + PC + PD < P'A + P'B + P'C + P'D.$$

So, point $P$ indeed minimizes the sum of the distances.

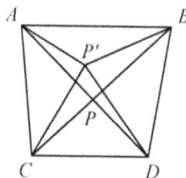

Fig. 9.5

Remark: "The shortest line connecting two points is the line segment" can be used not only in triangles but also in some shortest route problems, such as the following example.

**Example 4.** As shown in Figure 9.6, a little shepherd boy started from point $A$, drove the cattle to the river to drink water, and then went to point $B$. How to choose the drinking location so that the route the cattle took was the shortest?

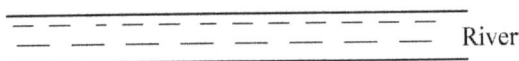

Fig. 9.6

**Solution** We use a "symmetrical" method to solve this problem.

Find the symmetric point $B'$ of $B$ with respect to the river bank. Connect $AB'$, and suppose that it intersects with the river bank at $C$. Point $C$ is the best position (as shown in Figure 9.7). That is, the shepherd boy should drive the cattle from $A$ to $C$ to drink water, and then go to point $B$, which is the shortest route. The reasons are as follows.

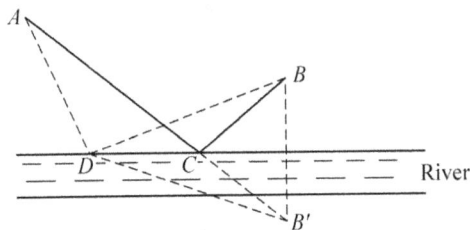

Fig. 9.7

If the drinking location is another point $D$, then the length of the route is $AD + BD$.

Connect $B'D$. As $B$ and $B'$ are symmetric about the river bank, we have

$$BC = B'C, BD = B'D,$$

then $AD + BD = AD + B'D$ and $AC + BC = AC + B'C = AB'$.

In $\triangle ADB'$, we have $AD + B'D > AB'$, and so $AD + BD > AC + BC$. That is, point $C$ is the desired drinking location.

Remark:   The essence of the solution is to use symmetry to convert the problem into one of finding the shortest route between points $A$ and $B'$. Since the line segment between two points is the shortest, we can easily find the shortest route.

**Example 5.** As shown in Figure 9.8, points $C$, $D$, and $E$ divide line segment $AB$ into four parts, and $AC : CD : DE : EB = 2 : 3 : 4 : 5$. Points $M$, $P$, $Q$, and $N$ are the midpoints of $AC$, $CD$, $DE$, and $EB$, respectively. If $MN = 21$, find the length of $PQ$.

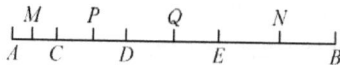

Fig. 9.8

**Solution**   Suppose $AC = 2k$. By the conditions, we have $CD = 3k$, $DE = 4k$, $EB = 5k$, and

$$MN = MC + CD + DE + EN$$
$$= \frac{1}{2}AC + CD + DE + \frac{1}{2}EB$$
$$= \frac{1}{2} \times 2k + 3k + 4k + \frac{1}{2} \times 5k$$
$$= \frac{21}{2}k,$$

so $\frac{21}{2}k = 21$, $k = 2$.

Now, we find

$$PQ = PD + DQ$$
$$= \frac{1}{2}CD + \frac{1}{2}DE$$
$$= \frac{1}{2} \times 3k + \frac{1}{2} \times 4k$$
$$= \frac{7}{2}k = 7.$$

**Answer**   The length of $PQ$ is 7.

Remark: Using algebra to solve geometric problems is also a common method. This is usually done when calculating the length or angle. The condition of this problem is a continued ratio, and the part corresponding to 1 should be assumed as $k$.

**Example 6.** As shown in Figure 9.9, $AB = 2BC$, $DA = \frac{3}{2}AB$, $M$ is the midpoint of $AD$, and $N$ is the midpoint of $AC$. Try to compare the lengths of $MN$ and $AB + NB$.

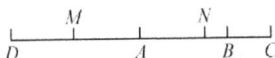

Fig. 9.9

**Solution** Suppose $BC = 2$, then $AB = 4$, $DA = 6$, and $AC = 4 + 2 = 6$.

$$2MN = 2MA + 2AN$$
$$= AD + AC$$
$$= 6 + 6 = 12.$$

So, $MN = 6$.

$$AB + NB = AB + (NC - BC)$$
$$= AB + \frac{1}{2}AC - \frac{1}{2}AB$$
$$= 4 + \frac{1}{2} \times 6 - \frac{1}{2} \times 4$$
$$= 5.$$

So, we have $MN > AB + NB$.

Remark: The line segments in this problem are all related to $BC$. Therefore, we use the length of $BC$ to represent the length of each line segment. This makes it easy to compare the lengths of the line segments. The length of $BC$ can be regarded as 1 (unit length) or any other number, and we use 2 (unit length) as the length of $BC$, mainly to avoid the occurrence of fractions. It is also fine to regard the length of $BC$ as $a$ (unit length). If you are interested, you can give it a try.

**Example 7.** As shown in Figure 9.10, $A$, $B$, and $C$ are three villages on a road. The distance between $A$ and $B$ is 100 km, while the distance between

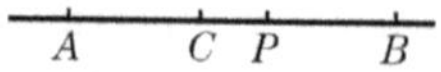

Fig. 9.10

$A$ and $C$ is 40 km. Locate a station $P$ between $A$ and $B$, and suppose that the distance between $P$ and $C$ is $x$ km.

(1) Use an algebraic formula involving $x$ to represent the sum of the distances from the station to the three villages.
(2) If the sum of the distances from the station to the three villages is 102 km, where is the station located?
(3) To minimize the sum of the distances from the station to the three villages, where should the station be located?

**Solution**

(1) The sum of the distances from the station $P$ to the three villages $A$, $B$, and $C$ is

$$PA + PB + PC = (PA + PB) + PC$$
$$= AB + PC$$
$$= AB + x$$
$$= (100 + x)(\text{km}).$$

(2) According to the condition, $100 + x = 102$, so $x = 2$, which is the distance from the station to village $C$ (there are two locations: on either side of village $C$).
(3) As $x \geqslant 0$, $100 + x$ reaches its minimum when $x = 0$. In this case, the station is located in village $C$. That is, to minimize the sum of the distances from the station to the three villages, the station should be located at village $C$.

**Example 8.** On the number line, $O$ is the origin. Points $A$ and $B$ represent the numbers $a$ and $b$, respectively. Find the number represented by the midpoint $C$ of the line segment $AB$ (in terms of $a$ and $b$).

**Analysis**  Discuss the given situation. Suppose $C$ represents the number $c$.

**Solution**  Without loss of generality, we may suppose that $A$ is on the left side of $B$. Then, there will be four cases:

(1) $A$ is on the right side of point $O$ (which is $a > 0$).
We have

$$OC = OA + AC,$$
$$OC = OB - CB,$$

where $OC$ represents the length of the line segment $OC$.
Add the above two formulas and use $AC = CB$ to find

$$OC = \frac{OA + OB}{2}. \qquad (9.2)$$

As $OA = a$, $OB = b$, and $OC = c$, (9.2) is simply

$$c = \frac{a + b}{2}. \qquad (9.3)$$

(2) $B$ is on the left side of point $O$ (which is $b < 0$).

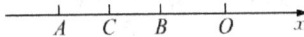

We have

$$CO = AO - AC,$$
$$CO = BO + CB.$$

Similarly, adding the above two formulas, we find

$$CO = \frac{AO + BO}{2}. \qquad (9.4)$$

As $AO = -a$, $BO = -b$, and $CO = -c$, substitute them in (9.4), and
we get (9.3).

(3) $O$ is between $A$ and $B$, and $C$ is on the right side of $O$ (which is $a < 0 < b$ and $c > 0$).

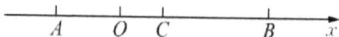

We have

$$OC = AC - AO,$$
$$OC = OB - CB,$$

Adding the above two formulas, we have

$$OC = \frac{OB - AO}{2}. \tag{9.5}$$

As $AO = -a$, $OB = b$, and $OC = c$, substitute them in (9.5), and we get (9.3).

(4) $O$ is between $A$ and $B$, and $C$ is on the left side of $O$ (which is $a < 0 < b$ and $c < 0$).

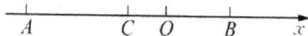

We have

$$CO = AO - AC,$$
$$CO = CB - OB.$$

Adding the above two formulas, we have

$$CO = \frac{AO - OB}{2}. \tag{9.6}$$

As $AO = -a$, $OB = b$, and $CO = -c$, substitute them in (9.6), and we find (9.3).

So, in all circumstances, we always have

$$c = \frac{a + b}{2}. \tag{9.7}$$

**Remark:** If a directed line segment is introduced (see "Reading"), the four formulas (9.2), (9.4), (9.5), and (9.6) can be unified into (9.2).

**Example 9.** There are 2009 different points on the line $\ell$. There are $\frac{2009 \times 2008}{2}$ line segments with these points as the endpoints. At least how many different midpoints do these line segments have?

**Analysis**   Use the conclusion of the previous problem.

**Solution** We first consider a special case where the 2009 points are on the number line and they represent $0, 1, \ldots, 2008$, respectively (that is, take $\ell$ as the number line and the leftmost point as the origin).

Now, for the point $\frac{s}{2}$ (that is, the point representing $\frac{s}{2}$, where $s$ is a natural number less than $2008 \times 2$), when $s$ is an even number $2k$, it is the midpoint of the line segment connecting $k - 1$ and $k + 1$, and when $s$ is an odd number $2k - 1$, it is the midpoint of the line segment connecting $k - 1$ and $k$. So, there are at least $2008 \times 2 - 1 = 4015$ midpoints.

On the other hand, the midpoints of any line segments whose endpoints are among the points $0, 1, \ldots, 2008$ are numbers in the form of $\frac{s}{2}$, and $s$ is a natural number smaller than $2 \times 2008$. Therefore, there are exactly 4015 midpoints.

For the general situation, these points can still be set on the number line, respectively representing the numbers $0 < a_1 < a_2 < \cdots < a_{2008}$. Then, all of these 4015 points,

$$\frac{a_1}{2} < \frac{a_2}{2} < \cdots < \frac{a_{2008}}{2} < \frac{a_1 + a_{2008}}{2} < \frac{a_2 + a_{2008}}{2} < \cdots < \frac{a_{2007} + a_{2008}}{2},$$

are midpoints of line segments: $\frac{a_k}{2}$ is the midpoint of $0$ and $a_k$ ($k = 1, 2, \ldots, 2008$); $\frac{a_k + a_{2008}}{2}$ is the midpoint of $a_k$ and $a_{2008}$ ($k = 1, 2, \ldots, 2007$). So, the number of midpoints is at least 4015.

Remark: Another proof is that there is a midpoint $\frac{a_1}{2}$ between the two points, $0 < a_1$. After adding a point $a_2 > a_1$, at least two midpoints $(\frac{a_1}{2} <) \frac{a_2}{2} < \frac{a_1 + a_2}{2}$ are added. By analogy, at least two midpoints will be added each time. When adding the last point $a_{2008}$, two midpoints $\frac{a_{2006} + a_{2008}}{2} < \frac{a_{2007} + a_{2008}}{2}$ (note that the existing rightmost midpoint $\frac{a_{2006} + a_{2007}}{2} < \frac{a_{2006} + a_{2008}}{2}$) will be added. So, there are at least $1 + 2 \times 2007 = 4015$ midpoints.

Generally, for $n$ ($n \geqslant 2$) points on a straight line, there are at least $2n - 3$ midpoints.

## Reading

---

### Line Segments on the Number Line

If the points $A$ and $B$ are on the number line and represent the numbers $a$ and $b$, respectively, then the length of the line segment $AB$ is $|b - a|$, regardless of the order of $A$ and $B$ and the location of the origin $O$ (as shown in Figure 9.11).

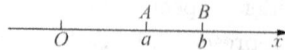

Fig. 9.11

We can also use $b - a$ as the length of $AB$ and $a - b$ as the length of $BA$ (the number represented by the second letter minus the number represented by the first letter). In this way, the length of the line segment can be positive or negative. The direction from $A$ to $B$ is consistent with the direction of the number line. That is, when $b > a$, the length of the line segment $AB$ is positive (the length of $BA$ is negative at this time); when $b < a$, the direction from $A$ to $B$ is opposite to the direction of the number line, and the length of the line segment $AB$ is negative (then the length of $BA$ is positive).

This kind of line segment whose length can be positive or negative is called a directed line segment (when $A$ and $B$ coincide, the length of the "line segment" $AB$ is 0).

There are many conclusions that can be drawn about directed line segments. For example, if $A$, $B$, and $C$ are three points on the number line, then

$$AB + BC + CA = 0, \tag{9.8}$$

$$AB \times OC + BC \times OA + CA \times OB = 0. \tag{9.9}$$

The proofs are also very simple. Suppose that the numbers represented by $A$, $B$, and $C$ are $a$, $b$, and $c$, respectively, then (9.8) is

$$(b - a) + (c - b) + (a - c) = 0, \tag{9.10}$$

and (9.9) is simply

$$(b - a)c + (c - b)a + (a - c)b = 0. \tag{9.11}$$

Here, (9.10) is obviously true. After expanding the left side of (9.11), the algebraic sum is 0. So, (9.11) is also obviously true.

---

**Exercises**

1. As shown in Figure 9.12, point $P$ is an arbitrary point in $\triangle ABC$. After extending $AP$, $BP$, and $CP$, they intersect with $BC$, $AC$, and $AB$ at

points $D$, $E$, and $F$, respectively. Prove that

$$AD + BE + CF > \frac{1}{2}(AB + BC + CA).$$

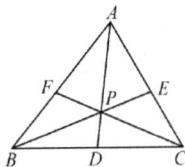

Fig. 9.12

2. As shown in Figure 9.13, point $P$ is an arbitrary point in $\triangle ABC$. Prove that

$$AB + BC + CA > PA + PB + PC.$$

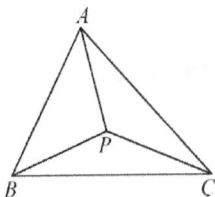

Fig. 9.13

3. As shown in Figure 9.14, point $A$ is a horse herding camp. Every day, the horse herders start from the camp, drive the horses to the river to drink water, then go to the grassland to graze, and then return to the camp. What is the shortest grazing route?

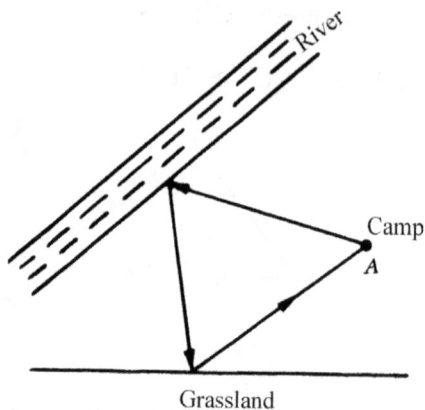

Fig. 9.14

4. As shown in Figure 9.15, there is a river between the two camps $A$ and $B$ (assuming that the river banks are parallel straight lines). How to build a bridge perpendicular to the river banks on the river and build roads from $A$ and $B$ to the bridge so that the total length of the road is the shortest?

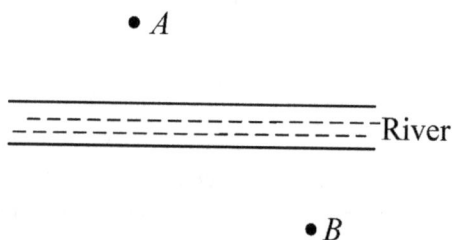

Fig. 9.15

5. As shown in Figure 9.16, given that the points $K$, $M$, $N$, and $H$ lie on the line segment $AB$, divide $AB$ into five small line segments of equal lengths.

(1) Write down all line segments in the figure whose endpoints are represented by letters.
(2) For which line segment(s), the midpoint is $M$?
(3) For which line segment(s), one trisection point is $N$?

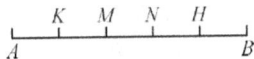

Fig. 9.16

6. As shown in Figure 9.17, the points $A$, $B$, $C$, $D$, and $E$ are on a straight line. $AB = CD$, and $E$ is the midpoint of $CB$. Is point $E$ the midpoint of $AD$? Why?

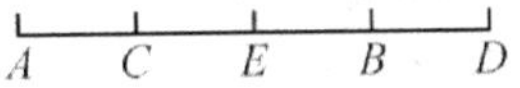

Fig. 9.17

7. As shown in Figure 9.18, the points $C$, $D$, and $E$ are on the line segment $AB$, such that $AC = CD = DE = EB$. The points $F$ and $G$ are the trisection points of the line segment $AB$. $AB = 12$ cm. Find the length of $CF + DF + EF$.

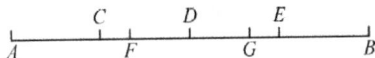

Fig. 9.18

8. As shown in Figure 9.19, given that $AB = 14$. There are four points $C$, $D$, $M$, and $N$ on the line segment $AB$ such that $AC : CD : DB = 1 : 2 : 4$, $AM = \frac{1}{2}AC$, and $DN = \frac{1}{4}DB$. Find the length of $MN$.

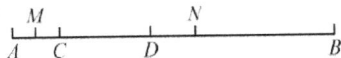

Fig. 9.19

9. As shown in Figure 9.20, we have that $AB = CD$ and $CB = \frac{1}{5}AB$. The points $E$ and $F$ are the midpoints of the line segments $AB$ and $CD$, respectively, and $EF = 12$ cm. Find the length of $AB$.

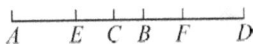

Fig. 9.20

10. As shown in Figure 9.21, the points $B$ and $C$ are two arbitrary points on the line segment $AD$. $B$ is between $A$ and $C$. The points $M$ and $N$

are the midpoints of $AB$ and $CD$, respectively. Given that $AD = a$ and $MN = b$, find the length of $BC$.

Fig. 9.21

11. Suppose that $P$ and $Q$ are two points on the line segment $AB$, $AB = 26$ cm, $AP = 14$ cm, and $PQ = 13$ cm. Find the length of $BQ$.
12. Suppose that $A$, $B$, and $C$ are collinear, with $AB = 16$. $D$ is the midpoint of $BC$ and $AD = 12$. Find the length of $BC$.

# Chapter 10

# Angles

The figure formed by two rays with a common endpoint is called an angle. Angles are also a basic figure of geometry. The basic concepts related to angles are round angles, straight angles, right angles, acute angles, obtuse angles, opposite angles, etc. In middle school geometry, angles generally refer to angles between $0°$ and $180°$.

**Example 1.** As shown in Figure 10.1, how many angles are there?

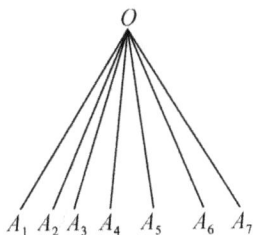

Fig. 10.1

**Solution** It can be seen from the figure that the largest angle $\angle A_1OA_7$ is divided into six parts by the five rays $OA_2$, $OA_3$, $OA_4$, $OA_5$, and $OA_6$. But if you think that there are only 6 angles in the figure, then you missed many angles. When counting the number of angles, be careful not to repeat or omit. We can count the angles with $OA_1$ as the left side first. From left to right, there are $\angle A_1OA_2$, $\angle A_1OA_3$, $\angle A_1OA_4$, $\angle A_1OA_5$, $\angle A_1OA_6$, and $\angle A_1OA_7$, totaling 6 angles. Then, count the angles with $OA_2$ as the left side. There are $A_2OA_3$, $\angle A_2OA_4$, $\angle A_2OA_5$, $\angle A_2OA_6$, and $\angle A_2OA_7$, totaling 5 angles. Likewise, with $OA_3$, $OA_4$, $OA_5$, and $OA_6$ as the left sides of the angles, there are 4, 3, 2, and 1 angles, respectively. So, the total number of angles in the figure is

$$6 + 5 + 4 + 3 + 2 + 1 = 21.$$

Remark: When counting angles in a figure, similarly to counting line segments, one should never repeat or omit. To this end, we must count them according to certain rules.

In addition, this problem can also be converted into a problem of counting line segments. As shown in Figure 10.2, draw a straight line that intersects $OA_1$, $OA_2$, ..., $OA_7$ at $A_1$, $A_2$, ..., $A_7$, respectively. Each angle corresponds to a certain line segment on $A_1A_7$. Conversely, each line segment on $A_1A_7$ corresponds to a certain angle. For example, $\angle A_3OA_5$ corresponds to the line segment $A_3A_5$, and the line segment $A_3A_5$ corresponds to $\angle A_3OA_5$. Therefore, to count the number of angles in the figure, we simply need to count the number of line segments on $A_1A_7$, which is

$$\frac{6 \times 7}{2} = 21.$$

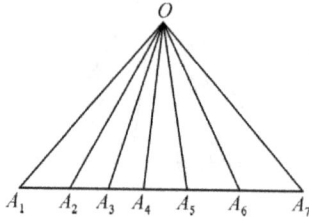

Fig. 10.2

Therefore, there are totally 21 angles in the figure.

In this way, the number of angles in Figure 10.3 can be calculated. A straight line intersects rays at $A_1$, $A_2$, ..., $A_{100}$. According to the formula for counting line segments, the total number of line segments on $A_1A_{100}$ is

$$\frac{99 \times 100}{2} = 4950.$$

Therefore, there are totally 4950 angles in the figure.

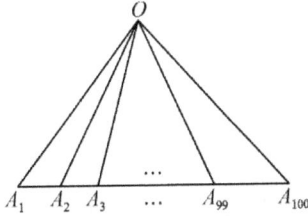

Fig. 10.3

**Example 2.** There is an angle such that twice its complementary angle and half of its supplementary angle are supplementary. Find the degree of this angle.

**Solution** Suppose that the angle is $x°$. Then, its complementary angle is $90° - x°$, and its supplementary angle is $180° - x°$. According to the condition, we have

$$2(90 - x) + \frac{1}{2}(180 - x) = 180,$$

On solving it, we have $x = 36$.

So, the angle is $36°$.

**Example 3.** As shown in Figure 10.4, there are six rays, $OA$, $OB$, $OC$, $OD$, $OE$, and $OF$, from the point $O$. $\angle AOB = 100°$, $OF$ bisects $\angle BOC$, $\angle AOE = \angle DOE$, and $\angle EOF = 140°$. Find the degree of $\angle COD$.

**Solution** As $\angle EOF + \angle BOF + \angle AOE + \angle AOB = 360°$, $\angle EOF = 140°$, and $\angle AOB = 100°$, we have

$$\angle BOF + \angle AOE = 360° - 140° - 100° = 120°.$$

According to the given conditions, $\angle BOF = \angle COF$, and $\angle AOE = \angle DOE$, we have

$$\angle EOD + \angle COF = \angle BOF + \angle AOE = 120°.$$

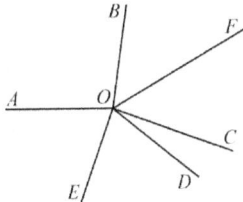

Fig. 10.4

Thus,

$$\angle COD = 140° - (\angle EOD + \angle COF) = 140° - 120° = 20°.$$

**Answer**   The degree of $\angle COD$ is $20°$.

Remark:   In this problem, we treat $\angle EOD + \angle COF$ as a whole. Its value can be obtained, and the problem can be solved (note: the degrees of $\angle EOD$ and $\angle COF$ cannot be obtained). When solving this problem, pay attention to the implicit condition: the sum of $\angle EOF$, $\angle BOF$, $\angle AOE$, and $\angle AOB$ is exactly one round angle, namely $360°$.

**Example 4.** Suppose that $\angle AOB = 40°$. There is a ray $OC$ from the point $O$, and $\angle AOC : \angle COB = 2 : 3$. Find the angle in degrees between $OC$ and the bisector of $\angle AOB$.

**Solution**   We discuss the problem in two cases, where $OC$ is either inside or outside $\angle AOB$ (as shown in Figure 10.5, where $OD$ is the bisector of $\angle AOB$).

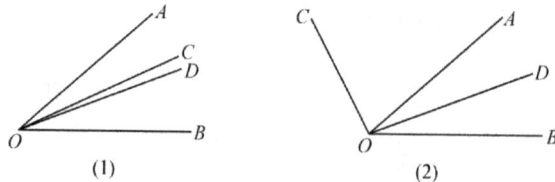

Fig. 10.5

(1) $OC$ is inside $\angle AOB$:
   As $\angle AOC : \angle COB = 2 : 3$ and $\angle AOB = 40°$, we have

$$\angle AOC = 40° \times \frac{2}{2+3} = 16°.$$

As $OD$ bisects $\angle AOB$, we have

$$\angle AOD = \frac{1}{2} \times 40° = 20°.$$

$$\angle COD = \angle AOD - \angle AOC = 20° - 16° = 4°.$$

(2) $OC$ is outside $\angle AOB$:

As $\angle AOC : \angle COB = 2 : 3$ and $\angle AOB = 40°$, we have

$$\angle COA = 40° \div (3 - 2) \times 2 = 80°.$$

As $\angle AOD = 20°$, we have

$$\angle COD = \angle COA + \angle AOD = 80° + 20° = 100°.$$

Remark: There are two cases in this problem. We should not miss either of them.

**Example 5.** Suppose that $\alpha$ and $\beta$ are both obtuse angles. Four people A, B, C, and D compute $\frac{1}{6}(\alpha + \beta)$, and the results are $28°$, $48°$, $88°$, and $60°$, respectively. Only one of them is correct. Which result is correct?

**Solution** As $\alpha$ and $\beta$ are obtuse angles, we have

$$90° < \alpha < 180°, 90° < \beta < 180°.$$

It follows that $180° < \alpha + \beta < 360°$ and $30° < \frac{1}{6}(\alpha + \beta) < 60°$.

That is, the value of $\frac{1}{6}(\alpha + \beta)$ should be between $30°$ and $60°$ (excluding $30°$ and $60°$). Checking the results of A, B, C, and D, only the result obtained by B is in this range. So, $48°$ is correct.

**Example 6.** As shown in Figure 10.6, $\angle A_1 O A_{11}$ is a straight angle, rays $O A_2, O A_3, \ldots, O A_{10}$ are all within this straight angle (in clockwise order), and $\angle A_3 O A_2 - \angle A_2 O A_1 = \angle A_4 O A_3 - \angle A_3 O A_2 = \angle A_5 O A_4 - \angle A_4 O A_3 = \cdots = \angle A_{11} O A_{10} - \angle A_{10} O A_9 = 2°$. Find the degree of $\angle A_{10} O A_{11}$.

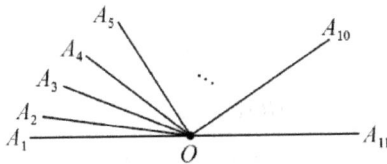

Fig. 10.6

**Solution**  Suppose the degrees of $\angle A_1OA_2$, $\angle A_2OA_3, \ldots, \angle A_{10}OA_{11}$ are $\alpha_1, \alpha_2, \ldots, \alpha_{10}$, respectively. Then, we have

$$\alpha_1 + \alpha_2 + \cdots + \alpha_{10} = 180°, \tag{10.1}$$

and

$$\alpha_9 = \alpha_{10} - 2°,$$

$$\alpha_8 = \alpha_9 - 2° = \alpha_{10} - 2 \times 2°,$$

$$\alpha_7 = \alpha_8 - 2° = \alpha_{10} - 3 \times 2°,$$

$$\cdots$$

$$\alpha_1 = \alpha_2 - 2° = \alpha_{10} - 9 \times 2°.$$

Substitute them in (10.1) to get

$$10\alpha_{10} - 2° \times (1 + 2 + \cdots + 9) = 180°,$$

$$\alpha_{10} = 27°.$$

That is, $\angle A_{10}OA_{11} = 27°$.

Remark:  The degrees of $\angle A_1OA_2$, $\angle A_2OA_3, \ldots, \angle A_{10}OA_{11}$ in this problem form an arithmetic sequence with a common difference of 2.

**Example 7.** If two angles are opposite to each other, do their bisectors lie on a straight line? Why?

**Solution**  They are on a straight line. As shown in Figure 10.7, lines $AB$ and $CD$ intersect at a point $O$. $\angle AOC$ and $\angle BOD$ form a pair of vertical angles. Suppose $OE$ and $OF$ are the bisectors of $\angle AOC$ and $\angle BOD$. respectively. Next, we prove that $OE$ and $OF$ are on a straight line.

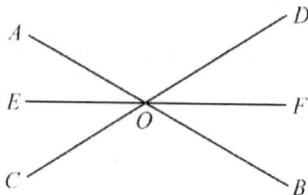

Fig. 10.7

As $\angle AOE = \frac{1}{2}\angle AOC$, $\angle BOF = \frac{1}{2}\angle BOD$, and $\angle AOC = \angle BOD$, we have $\angle AOE = \angle BOF$.

Also, as $\angle BOF + \angle FOD + \angle DOA = 180°$, $\angle AOE + \angle FOD + \angle DOA = 180°$.

That is, $\angle EOF = 180°$.
So, $OE$ and $OF$ are on the same line.

Remark: The method to prove that two rays $OE$ and $OF$ with the same endpoint form a straight line is to prove that $\angle EOF = 180°$, that is, $\angle EOF$ is a straight angle. This is justified by argument and calculation, not by sight.

**Example 8.** From 2 o'clock to 4 o'clock on the clock face, how many times do the hour hand and the minute hand form an angle of 60°? When?

**Solution** There are four times that the angle between the hour hand and the minute hand is 60°.

(1) The first time is at exactly two o'clock. Obviously, at this time, the hour hand and the minute hand form an angle of 60°. (Because the central angle between every two adjacent numbers on the clock face is $\frac{360°}{12} = 30°$.)

(2) The angle between the hour hand and the minute hand is 60° for the second time, as shown in Figure 10.8(1). Suppose it is at 2 o'clock $x$ minute. At this time, the minute hand moves $x$ minutes, that is, $x$ minute marks, and the hour hand moves $\frac{x}{12}$ minute marks (in 1 h, the minute hand finishes one rotation, or 60 min marks, and the hour hand finishes 5 min marks. So, the speed of the hour hand is $\frac{1}{12}$ of the minute hand). Since each minute interval is $\frac{360°}{60} = 6°$, the minute hand must exceed the hour hand by 10 min marks to form 60°. So,

$$x = \frac{x}{12} + 10 + 10,$$

On solving it, we have $x = 21\frac{9}{11}$; that is, at 2 o'clock $21\frac{9}{11}$ min, the angle between the hour hand and the minute hand is 60°.

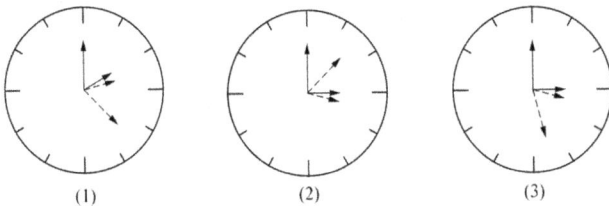

(1)      (2)      (3)

Fig. 10.8

(3) The angle between the hour hand and the minute hand is 60° for the third time, as shown in Figure 10.8(2). Suppose it is 3 o'clock and $y$ minute. Then,

$$y + 10 = \frac{y}{12} + 15.$$

On solving it, we have $y = 5\frac{5}{11}$; that is, at 3 o'clock $5\frac{5}{11}$ min, the angle between the hour hand and the minute hand is 60°.

(4) For the fourth time, the angle between the hour hand and the minute hand is 60°, as shown in Figure 10.8(3). Suppose it is 3 o'clock and $z$ minutes, then

$$z = 15 + \frac{z}{12} + 10.$$

After solving it, we have $z = 27\frac{3}{11}$; that is, at 3 o'clock $27\frac{3}{11}$ min, the angle between the hour hand and the minute hand is 60°.

## Reading

---

### Angles produced by Rotation

If a ray $OA$ is rotated around from the point $O$ to $OB$, then an angle is produced, that is, $\angle AOB$.

Rotating $OA$ to different positions of $OB$ produces different angles.

We can also distinguish two kinds of rotations: one is clockwise rotation, which is in the same direction as the hands on the clock; the other is counterclockwise rotation, which is opposite to the rotation direction of the hands on the clock. Figure 10.9 shows a counterclockwise rotation.

When driving a car, turning the steering wheel to the left (counterclockwise) is completely different from turning the steering wheel to the right (clockwise).

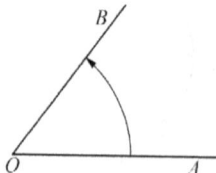

Fig. 10.9

We can regard an angle produced by counterclockwise rotation as positive and an angle produced by clockwise rotation as negative. In this way, an angle has a sign.

___

**Exercises**

1. Determine whether the following propositions are true or false:
   (1) Subtracting an acute angle from an obtuse angle, the difference must be an acute angle.                                    (   )
   (2) Subtracting an acute angle from a right angle, the difference must be an acute angle.                                    (   )
   (3) If two supplementary angles have a common side, then the angle formed by the bisectors of these two angles must be a right angle.                                                (   )
   (4) Two equal angles must be opposite angles.            (   )
   (5) If two equal angles have a common side, then the other sides must be on the same straight line.                            (   )

2. An angle is $20°$ greater than its complementary angle. What is the degree of the supplementary angle of its complementary angle?

3. For an angle, the sum of its supplementary angle and complementary angle is $60°$ greater than the difference between its supplementary angle and complementary angle. Find the degree of the complementary angle of this angle.

4. As shown in Figure 10.10, $\angle AOB = 90°$, $\angle AOC$ is an acute angle, $ON$ bisects $\angle AOC$, and $OM$ bisects $\angle BOC$. Find the degree of $\angle MON$.

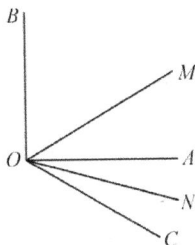

Fig. 10.10

5. As shown in Figure 10.11, $\angle AOB$ and $\angle COD$ are the complementary angles of $\angle BOC$, and $OE$ and $OF$ are the bisectors of $\angle AOB$

and $\angle COD$, respectively. If $\angle BOC = 50°$, find the degrees of $\angle AOE$, $\angle AOD$, and $\angle EOF$.

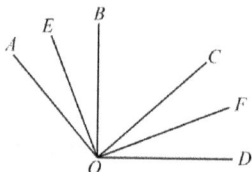

Fig. 10.11

6. Given an angle, the sum of its supplementary angle and three times its complementary angle is equal to $\frac{11}{12}$ of the round angle. Find the degree of this angle.

7. As shown in Figure 10.12, the lines $AB$ and $CD$ intersect at $O$, $OE$ bisects $\angle AOC$, and $\angle BOC - \angle BOD = 20°$. Find the degree of $\angle BOE$.

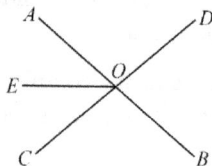

Fig. 10.12

8. As shown in Figure 10.13, how many angles are there in total?

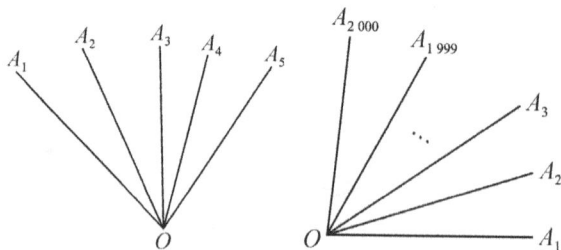

Fig. 10.13

9. As shown in Figure 10.14, $O$ is a point on the line $AB$, and $OD$ and $OE$ are the bisectors of $\angle AOC$ and $\angle BOC$, respectively. Prove that $\angle DOE = 90°$.

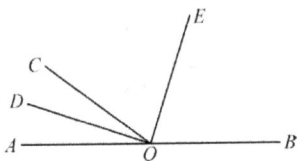

Fig. 10.14

10. Prove that twice the complementary angle of half an acute angle, minus the supplementary angle of twice this acute angle, is equal to the original angle.

11. As shown in Figure 10.15, it is a $3 \times 3$ grid. Find the degree of $\angle 1 + \angle 2 + \angle 3 + \angle 4 + \angle 5 + \angle 6 + \angle 7 + \angle 8 + \angle 9$.

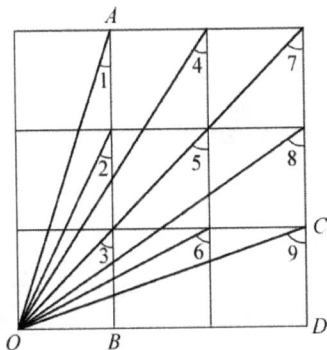

Fig. 10.15

12. As shown in Figure 10.16, given that $\angle A_3OA_2 - \angle A_2OA_1 = \angle A_4OA_3 - \angle A_3OA_2 = \angle A_5OA_4 - \angle A_4OA_3 = \cdots = \angle A_8OA_7 - \angle A_7OA_6 = \angle A_1OA_8 - \angle A_8OA_7 = 4°$, find the degree of $\angle A_2OA_3$.

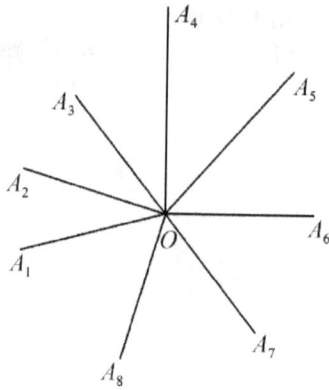

Fig. 10.16

13. At 8:20, what is the degree of the angle formed by the hour hand and the minute hand?

14. Between 3 o'clock and 4 o'clock (including 3 o'clock and 4 o'clock), when do the hour hand and the minute hand form an angle of 90°?

15. Two equal angles have a common vertex and a common side, and the angle formed by the other two sides is a right angle. Find the degree of these two angles.

# Chapter 11

# Sum of the Interior Angles of a Triangle

It is known that the sum of the interior angles of a triangle is equal to $180°$. This is a property of all triangles. It indicates that an exterior angle of a triangle is equal to the sum of the two interior angles that are not adjacent to it.

**Example 1.** As shown in Figure 11.1, the quadrilateral $ABCD$ is an arbitrary quadrilateral. Find the sum of its interior angles.

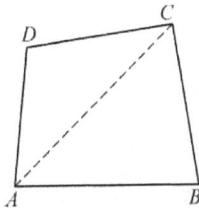

Fig. 11.1

**Solution** Connect $AC$. Then, the quadrilateral $ABCD$ is divided into two triangles, that is, $\triangle ABC$ and $\triangle ACD$. Therefore, the sum of the interior angles of the quadrilateral $ABCD$ is equal to the sum of the interior angles of the two triangles. The sum of the interior angles of one triangle is $180°$, and so the sum of the interior angles of the quadrilateral is

$$180° \times 2 = 360°.$$

Remark:   In this problem, we divide a quadrilateral into two triangles by
connecting $AC$. This method can be generalized. That is to say, in general,
to find the sum of the interior angles of a convex $n$-gon, we can start with
one of its vertices $A_1$ and connect $A_1A_3$, $A_1A_4$, ..., $A_1A_{n-1}$, which divides
this $n$-gon into $n-2$ triangles. Therefore, the sum of the interior angles of
this $n$-gon is $180° \times (n-2)$ (as shown in Figure 11.2).

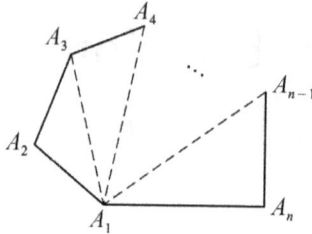

Fig. 11.2

This can be used as a formula. For example, the sum of the interior
angles of a 100-gon is

$$180° \times (100 - 2) = 17640°.$$

**Example 2.** Prove that the sum of the exterior angles of a triangle is equal
to 360°. Generally, the sum of the exterior angles of a convex $n$-gon is equal
to 360°.

**Proof**   As shown in Figure 11.3, in $\triangle ABC$, $\angle 1$, $\angle 2$, and $\angle 3$ are interior
angles. $\angle 4$, $\angle 5$, and $\angle 6$ are exterior angles.

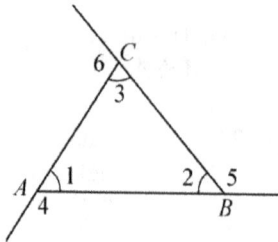

Fig. 11.3

Then, we have

$$\angle 1 + \angle 4 = 180°,$$

$$\angle 2 + \angle 5 = 180°,$$

$$\angle 3 + \angle 6 = 180°.$$

So,

$$\angle 4 + \angle 5 + \angle 6 = 3 \times 180° - (\angle 1 + \angle 2 + \angle 3)$$

$$= 3 \times 180° - 180°$$

$$= 360°.$$

For the same reason, if $\angle\alpha_1, \angle\alpha_2, \ldots, \angle\alpha_n$ are $n$ interior angles of an $n$-gon and $\angle\beta_1, \angle\beta_2, \ldots, \angle\beta_n$ are its $n$ exterior angles, then we have

$$\angle\alpha_1 + \angle\beta_1 = 180°,$$

$$\angle\alpha_2 + \angle\beta_2 = 180°,$$

$$\cdots$$

$$\angle\alpha_n + \angle\beta_n = 180°.$$

So, we have

$$\angle\beta_1 + \angle\beta_2 + \cdots + \angle\beta_n = n \times 180° - (\angle\alpha_1 + \angle\alpha_2 + \cdots + \angle\alpha_n)$$

$$= n \times 180° - (n-2) \times 180°$$

$$= 360°.$$

**Example 3.** Given that the second interior angle of a quadrilateral is three times the first interior angle, the third interior angle is half of the second interior angle, and the fourth interior angle is 10° greater than the third interior angle, find the degree of its first interior angle.

**Solution** Suppose the first interior angle is $x$, then its second interior angle is $3x$, its third interior angle is $\frac{3}{2}x$, and its fourth interior angle is $\frac{3}{2}x + 10°$. As the sum of the interior angles of a quadrilateral is 360°, we have

$$x + 3x + \frac{3}{2}x + \left(\frac{3}{2}x + 10°\right) = 360°.$$

On solving it, we have $x = 50°$.

**Answer** Its first interior angle is 50°.

**Example 4.** If the largest angle of a triangle is four times the smallest angle, then find the range of its smallest angle.

**Solution**   Suppose $\angle A$ is the smallest angle, $\angle C$ is the largest angle, and $\angle B$ is the middle one, then $\angle A \leqslant \angle B \leqslant \angle C$ and $\angle C = 4\angle A$.
    By

$$\begin{cases} \angle A + \angle B + \angle C = 180°, \\ \angle C = 4\angle A, \\ \angle A \leqslant \angle B \leqslant \angle C, \end{cases}$$

    We have $\angle A + \angle A + 4\angle A \leqslant 180°$ (as $\angle A \leqslant \angle B$ and $4\angle A = \angle C$), that is, $\angle A \leqslant 30°$.

    Also, we have $\angle A + 4\angle A + 4\angle A \geqslant 180°$ (as $4\angle A = \angle C \geqslant \angle B$), that is, $\angle A \geqslant 20°$.

    So, the range of the smallest angle is $20° \leqslant \angle A \leqslant 30°$.

**Example 5.** As shown in Figure 11.4, in $\triangle ABC$, $BD$ is the bisector of $\angle ABC$, and $CD$ is the bisector of the exterior angle $\angle ACE$.

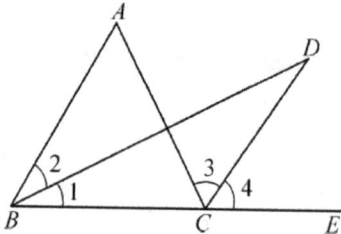

Fig. 11.4

    Prove that $\angle D = \frac{1}{2}\angle A$.

**Proof**   According to the property of the exterior angles of a triangle, we have $\angle 3 + \angle 4 = \angle 1 + \angle 2 + \angle A$.

    As $BD$ and $CD$ are the bisectors of $\angle ABC$ and $\angle ACE$, respectively, $\angle 1 = \angle 2$ and $\angle 3 = \angle 4$.

    Thus, $2\angle 4 = 2\angle 1 + \angle A$, that is,

$$\angle 4 = \angle 1 + \frac{1}{2}\angle A. \tag{11.1}$$

    In $\triangle BCD$, $\angle 4$ is an exterior angle, so we have

$$\angle 4 = \angle 1 + \angle D. \tag{11.2}$$

    By (11.1) and (11.2), we have $\angle D = \frac{1}{2}\angle A$.

**Example 6.** As shown in Figure 11.5, in the seven-star $ABCDEFG$, find the degree of $\angle A + \angle B + \angle C + \angle D + \angle E + \angle F + \angle G$.

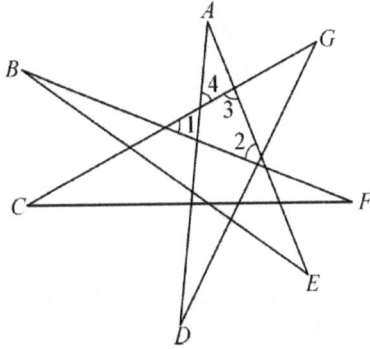

Fig. 11.5

**Solution** By the property of the exterior angles of a triangle, we have

$$\angle 1 = \angle C + \angle F,$$
$$\angle 2 = \angle B + \angle E,$$
$$\angle 4 = \angle D + \angle G,$$
$$\angle 3 = \angle 4 + \angle A = \angle D + \angle G + \angle A.$$

So, we have

$$\angle A + \angle B + \angle C + \angle D + \angle E + \angle F + \angle G$$
$$= \angle 1 + \angle 2 + \angle 3$$
$$= 180°.$$

**Remark:** In this problem, the seven requested angles are very scattered, and it is difficult to find their sum directly. Therefore, we use the property of the exterior angles of a triangle to gather them into a single triangle to solve the problem.

**Example 7.** As shown in Figure 11.6, $D$ is a point inside $\triangle ABC$. Prove that

$$\angle BDC = \angle A + \angle ABD + \angle ACD.$$

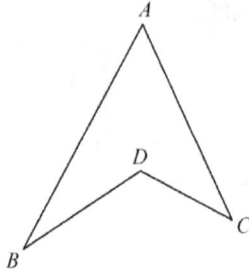

Fig. 11.6

**Proof** As shown in Figure 11.7, extend $BD$ to intersect $AC$ at point $E$. As $\angle BDC$ is an exterior angle of $\triangle CDE$, we have

$$\angle BDC = \angle DEC + \angle ACD.$$

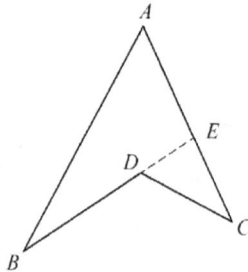

Fig. 11.7

Also, as $\angle DEC$ is an exterior angle of $\triangle AEB$, we have

$$\angle DEC = \angle A + \angle ABD.$$

By the above two formulas, we have

$$\angle BDC = \angle A + \angle ABD + \angle ACD.$$

Remark: The conclusion of this problem is frequently used in many problems.

Note that $D$ must be inside $\triangle ABC$ (that is, quadrilateral $ABDC$ is a concave quadrilateral, and it is concave at point $D$). Otherwise, the conclusion does not hold.

**Example 8.** As shown in Figure 11.8, $BE$ bisects $\angle ABD$ and $CF$ bisects $\angle ACD$. $BE$ and $CF$ intersect at point $G$. If $\angle BDC = 140°$ and $\angle BGC = 100°$, then find the degree of $\angle A$.

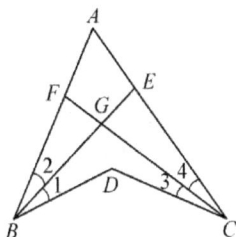

Fig. 11.8

**Solution**  From the previous example, we have

$$\angle BGC = \angle A + \angle 2 + \angle 4, \qquad (11.3)$$

$$\angle BDC = \angle A + (\angle 1 + \angle 2) + (\angle 3 + \angle 4). \qquad (11.4)$$

Also, as $BE$ bisects $\angle ABD$ and $CF$ bisects $\angle ACD$, we have $\angle 1 = \angle 2$ and $\angle 3 = \angle 4$.

So, (11.4) is simply

$$\angle BDC = \angle A + 2(\angle 2 + \angle 4). \qquad (11.5)$$

By (11.3) $\times$ 2 $-$ (11.5), we have

$$\angle A = 2\angle BGC - \angle BDC = 2 \times 100° - 140° = 60°.$$

**Example 9.** As shown in Figure 11.9, in $\triangle ABC$, $AD$ is the bisector of $\angle BAC$, and $CE$ is perpendicular to $AD$ and intersects $AD$ at $E$. Prove that $\angle ACE > \angle B$.

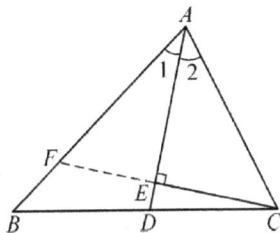

Fig. 11.9

**Proof** Extend $CE$ to intersect $AB$ at point $F$. As $AD$ is the bisector of $\angle BAC$, $\angle 1 = \angle 2$. Also, as $CE$ is perpendicular to $AD$, $\angle AEC = \angle AEF = 90°$.

In $\triangle AEF$, $\angle AFC = 180° - (\angle 1 + \angle AEF)$.
In $\triangle AEC$, $\angle ACE = 180° - (\angle 2 + \angle AEC)$.
So, $\angle ACE = \angle AFC$.
As $\angle AFC$ is an exterior angle of $\triangle BCF$, we have

$$\angle AFC = \angle B + \angle BCF > \angle B.$$

Thus, $\angle ACE > \angle B$.

Remark:   Using the given conditions, we construct $\angle AFC$ as a bridge: on the one hand, it is equal to $\angle ACE$; on the other hand, it is an exterior angle of $\triangle BFC$, which must be larger than any non-adjacent interior angle. This solves the problem.

**Example 10.** As shown in Figure 11.10, in $\triangle ABC$, $D$ and $E$ are points on the side $BC$, $\angle BDA = \angle BAD$, $\angle CEA = \angle CAE$, and $\angle DAE = \frac{1}{3}\angle BAC$.
    Find the degree of $\angle BAC$.

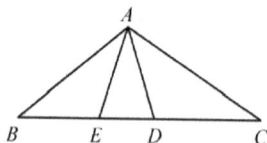

Fig. 11.10

**Solution**   Suppose $\angle BAE$, $\angle EAD$, and $\angle DAC$ are $\alpha$, $\beta$, and $\gamma$, respectively. Then, we have

$$\beta = \frac{1}{3}(\alpha + \beta + \gamma).$$

That is,

$$2\beta = \gamma + \alpha. \tag{11.6}$$

Also, we find

$$\angle BDA = \angle BAD = \alpha + \beta, \tag{11.7}$$

$$\angle CEA = \angle CAE = \beta + \gamma. \tag{11.8}$$

In $\triangle AED$, the sum of the interior angles is $180°$, so by (11.7) and (11.8), we have

$$(\alpha + \beta) + (\beta + \gamma) + \beta = 180°. \tag{11.9}$$

Combining it with (11.6), we get

$$5\beta = 180°,$$
$$\beta = 36°,$$
$$\angle BAC = 3\beta = 3 \times 36° = 108°.$$

**Remark:** In this problem, we apply the given conditions and the fact that the sum of the interior angles of a triangle is 180° to establish some equations to solve the problem.

## Reading

---

### Equivalent Propositions

Consider two sentences: "$x = 1$" and "$x + 2 = 3$." If the first sentence is correct, then the second sentence is also correct. Conversely, if the second sentence is correct, then the first sentence is also correct. Such two sentences are called equivalent propositions.

There are many equivalent propositions in mathematics. For example:

**Proposition 1.** The sum of the interior angles of a triangle is equal to a straight angle;
and

**Proposition 2.** An exterior angle of a triangle is equal to the sum of its two non-adjacent interior angles.

They are a pair of equivalent propositions.

If two propositions are equivalent, then as long as one of them is proved to be true, the other must also be true. Conversely, if one of them is proved to be false, the other must also be false.

---

## Exercises

1. Compute the sum of the interior and exterior angles of a convex 10-gon.
2. Given that one interior angle of a quadrilateral is 56°, the second interior angle is twice as large, and the third interior angle is 10° smaller than the second interior angle. Find the degree of the fourth interior angle.

3. As shown in Figure 11.11, $\angle A = 80°$, and the bisectors of $\angle ABC$ and the exterior angle of $\angle ACB$ intersect at $D$. Find the size of $\angle D$.

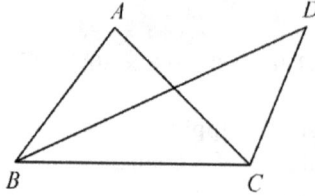

Fig. 11.11

4. As shown in Figure 11.12, find the degree of $\angle A + \angle B + \angle C + \angle D + \angle E$.

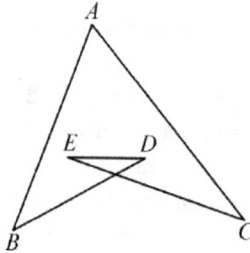

Fig. 11.12

5. As shown in Figure 11.13, find the degree of $\angle A + \angle B + \angle C + \angle D + \angle E$.

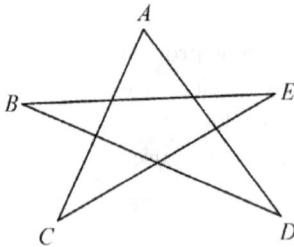

Fig. 11.13

6. As shown in Figure 11.14, find the degree of $\angle A + \angle B + \angle C + \angle D + \angle E + \angle F$.

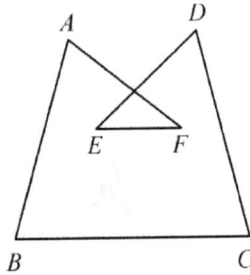

Fig. 11.14

7. As shown in Figure 11.15, a straight line intersects the sides and extended sides of $\triangle ABC$ at $D$, $E$, and $F$. $\angle A = 20°$, $\angle CED = 100°$, and $\angle ADF = 35°$. Find the degree of $\angle B$.

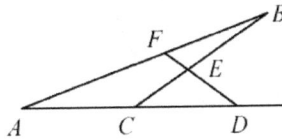

Fig. 11.15

8. As shown in Figure 11.16, in $\triangle ABC$, $\angle ABC = \angle C$, and $BD \perp AC$ meets $AC$ at $D$. Prove that $\angle DBC = \frac{1}{2}\angle A$.

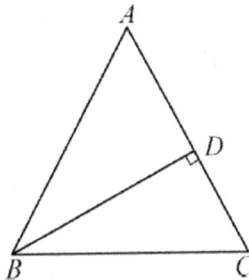

Fig. 11.16

9. As shown in Figure 11.17, in $\triangle ABC$, $AD$ is the bisector of $\angle BAC$ and $BE \perp AD$ with foot $E$ on the extension line of $AD$. Prove that $\angle ABE < \angle ACB$.

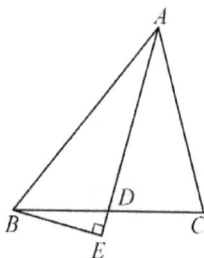

Fig. 11.17

10. As shown in Figure 11.18, the bisector of the exterior angle of $\angle C$ of $\triangle ABC$ intersects the extension line of $BA$ at $D$. Prove that $\angle BAC > \angle B$.

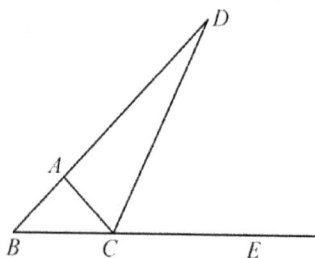

Fig. 11.18

11. As shown in Figure 11.19, $AB$ and $CD$ intersect at $E$. $CF$ and $BF$ are the bisectors of $\angle ACD$ and $\angle ABD$, respectively, and they intersect at $F$. Prove that $\angle F = \frac{1}{2}(\angle A + \angle D)$.

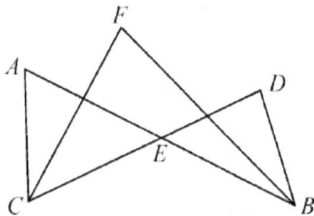

Fig. 11.19

12. As shown in Figure 11.20, $P$ is a point inside $\triangle ABC$. Prove that $\angle BPC > \angle A$.

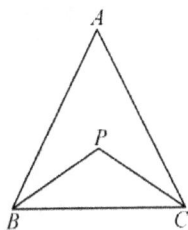

Fig. 11.20

13. There is a pentagon. Its first interior angle is $20°$ smaller than its second interior angle, its second interior angle is $20°$ smaller than its third interior angle, ..., its fourth interior angle is $20°$ smaller than its fifth interior angle. Find the degree of its third interior angle.

14. The three interior angles $\angle A$, $\angle B$, and $\angle C$ of $\triangle ABC$ satisfy the following conditions: $3\angle A > 5\angle B$ and $3\angle C \leqslant 2\angle B$.

   (1) Try to find two groups of $\angle A$, $\angle B$, and $\angle C$ that meet the conditions.

   (2) What kind of triangles satisfy the conditions? Why?

# Chapter 12

# Parallel Lines

Two non-overlapping straight lines on a plane have only two positional relationships: intersecting or parallel.

Through a point outside a straight line $a$, there can be only one straight line parallel to $a$. This is called the parallel postulate.

**Example 1.** As shown in Figure 12.1, the straight lines $a$, $b$, and $c$ are in the same plane, $a//b$, and $a$ and $c$ intersect at point $P$. Try to explain that $b$ and $c$ must also intersect.

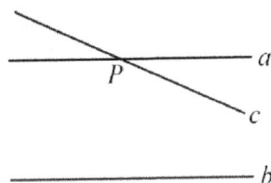

Fig. 12.1

**Solution**    Suppose that $b$ and $c$ do not intersect, then $b$ and $c$ are parallel.

The point $P$ is on the line $a$, and $a$ is parallel to $b$, so $P$ is not on the line $b$.

Through a point $P$ outside the straight line $b$, there are two straight lines $a$ and $c$ parallel to $b$. This contradicts the parallel postulate.

**Remark:** The method used in this problem is called "proof by contradiction." It is a common method in mathematical proofs.

When applying the method of proof by contradiction, we first suppose that the conclusion to be proved is not true, that is, the opposite conclusion is true. Then, we use the given conditions and assumptions to derive contradictions so as to show that the opposite conclusion is not true and the original conclusion is true.

To determine whether two straight lines are parallel, we usually use the following two theorems:

(1) Two straight lines are parallel if they are parallel to a third straight line.

(2) Two straight lines are parallel if they are perpendicular to a third straight line.

**Example 2.** As shown in Figure 12.2, $AB//CD$, $AB//EF$, $EG$ bisects $\angle BED$, $\angle B = 45°$, and $\angle D = 30°$. Find the degree of $\angle GEF$.

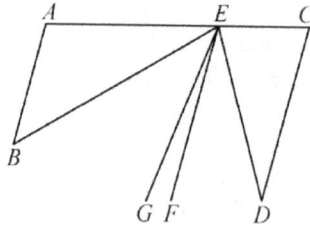

Fig. 12.2

**Solution** As $AB//CD$ and $AB//EF$, we have

$$EF//CD,$$

and so the interior alternate angle $\angle FED = \angle D = 30°$.

Also, as $AB//EF$, the interior alternate angle

$$\angle FEB = \angle B = 45°,$$

and so $\angle BED = \angle FED + \angle FEB = 75°$.

As $EG$ bisects $\angle BED$, we have

$$\angle BEG = \frac{1}{2}\angle BED = 37.5°.$$

Thus, $\angle GEF = \angle BEF - \angle BEG = 45° - 37.5° = 7.5°$.

So, $\angle GEF$ is $7.5°$.

**Example 3.** As shown in Figure 12.3, $AB//CD$, $E$ is between the lines $AB$ and $CD$, $\angle B = 40°$, and $\angle D = 20°$. Find the degree of $\angle BED$.

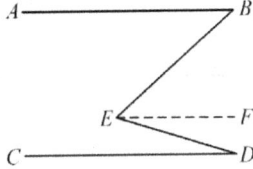

Fig. 12.3

**Solution**   Construct $EF//AB$ through $E$ as shown. Then, $EF//CD$, so we have

$$\angle FEB = \angle B = 40°, \angle DEF = \angle D = 20°,$$

and so $\angle BED = \angle BEF + \angle DEF = 40° + 20° = 60°$.

Remark:   Although this problem is very easy, it represents a typical class of problems, such as the following (as shown in Figures 12.4(1)–(4)). Suppose $AB//CD$:

(1) If $E$ is outside $AB$ and $CD$, similarly, we can construct $EF//AB$ through $E$, then $\angle BED = \angle FED - \angle FEB = \angle D - \angle B$.
(2) If point $E$ is below $CD$, for the same reason, $\angle BED = \angle B - \angle D$.
(3) By drawing the auxiliary lines shown in the figure, it is not difficult to prove that $\angle B + \angle BEF + \angle DFE + \angle D = 540°$.
(4) It can be proved by the same method that $\angle B + \angle EFG + \angle D = \angle BEF + \angle DGF$.

The key idea is to construct parallel lines of $AB$ through the points and use the properties of parallel lines to find the relationships between the angles.

We know that

if the straight lines $a$ and $b$ intersect $c$ and the corresponding angles or interior alternate angles are equal, then $a//b$ (the judgment theorem of parallel lines).

if the straight lines $a$ and $b$ intersect $c$ and $a//b$, then the corresponding angles are equal, and the interior alternate angles are also equal (the property theorem of parallel lines).

**Example 4.** As shown in Figure 12.5, $E$ is a point on $DF$, $B$ is a point on $AC$, $\angle 1 = \angle 2$, and $\angle C = \angle D$. Prove that $\angle A = \angle F$.

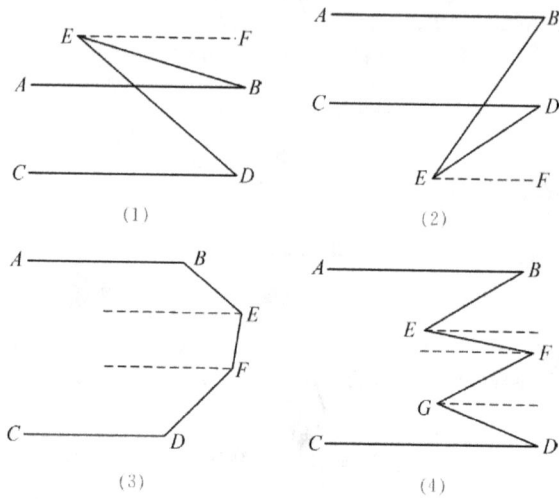

Fig. 12.4

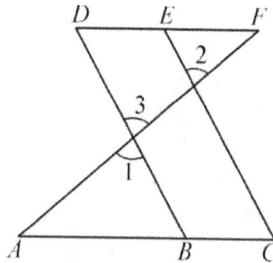

Fig. 12.5

**Proof**   As $\angle 1 = \angle 2$ and $\angle 1 = \angle 3$, we have

$$\angle 3 = \angle 2,$$

and so $BD//CE$, $\angle C = \angle ABD$.

Also, as $\angle C = \angle D$, we have $\angle ABD = \angle D$.

Thus, $AB//DF$, $\angle A = \angle F$.

Remark: We use the property and judgment theorems of parallel lines to solve the problem.

**Example 5.** As shown in Figure 12.6, $AB//CD$, $AE$ bisects $\angle BAC$, and $CE$ bisects $\angle ACD$. Prove that $AE \perp CE$.

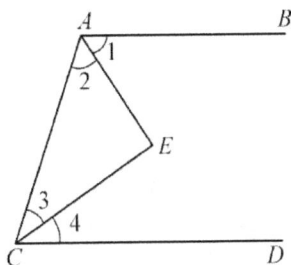

Fig. 12.6

**Proof** As $AB//CD$, we have

$$\angle BAC + \angle ACD = 180°.$$

Also, as $AE$ bisects $\angle BAC$ and $CE$ bisects $\angle ACD$, we get

$$\angle 2 = \frac{1}{2}\angle BAC,$$

$$\angle 3 = \frac{1}{2}\angle ACD.$$

So, we find

$$\angle 2 + \angle 3 = \frac{1}{2}(\angle BAC + \angle ACD)$$

$$= 90°,$$

$$\angle E = 180° - (\angle 2 + \angle 3)$$

$$= 90°,$$

that is, $AE \perp CE$.

**Example 6.** Seven straight lines are in a plane, with no two being parallel. Prove that among all the angles of intersections, there is at least one angle less than $26°$.

**Proof** Take any point $O$ on the plane, and through the point $O$, draw parallel lines $\ell_1'$, $\ell_2'$, $\ell_3'$, $\ell_4'$, $\ell_5'$, $\ell_6'$, and $\ell_7'$ of these seven straight lines, respectively. According to the property of parallel lines, the angles formed by $\ell_1'$, $\ell_2'$, $\ell_3'$, $\ell_4'$, $\ell_5'$, $\ell_6'$, and $\ell_7'$ are equal to the angles formed by the original seven straight lines $\ell_1$, $\ell_2$, ..., $\ell_7$, respectively.

Therefore, we can examine the angles formed by the adjacent lines, say $\ell_1'$ and $\ell_2'$, $\ell_2'$ and $\ell_3'$, ..., $\ell_7'$ and $\ell_1'$. From Figure 12.7, it is not difficult to find that the sum of these seven angles is a straight angle, that is, $180°$.

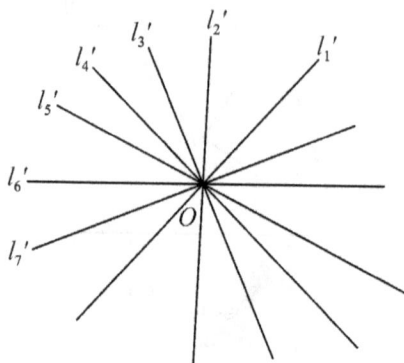

Fig. 12.7

Suppose that none of these seven angles is less than 26°, then the sum of these seven angles is at least $26° \times 7 = 182°$. This is impossible. Therefore, at least one of these seven angles is less than 26°. Without loss of generality, we suppose that the angle between $\ell_1'$ and $\ell_2'$ is less than 26°, then the angle formed between the original straight lines $\ell_1$ and $\ell_2$ is less than 26°.

Remark: Through translation, we can move the previously scattered angles to a single point for consideration.

## Reading

### Parallel Postulate

"Through a point $A$ outside the line $a$, at most one line $b$ parallel to $a$ can be drawn."

This proposition is called the parallel postulate (axiom), also known as the fifth postulate of Euclid. It is equivalent to "the sum of the interior angles of a triangle is equal to 180°."

Many people tried to prove the parallel postulate, but they all failed. Some outstanding mathematicians, such as Lobachevsky, Bolyai, and Gauss, thought that it might be possible to suppose that the parallel postulate is not true, that is, "through a point $A$ outside the line $a$, you can make two straight lines $b$ parallel to $a$." In this way, a new geometry was created, which is called non-Euclidean geometry.

## Exercises

1. As shown in Figure 12.8, $AB//CD$, $BE$, and $DE$ intersect at $E$. Prove that $\angle ABE = \angle D + \angle E$.

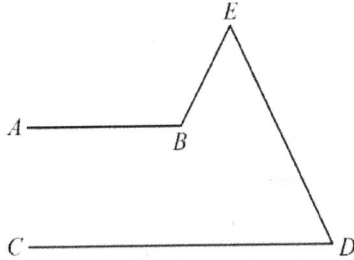

Fig. 12.8

2. As shown in Figure 12.9, $AB//CD$. Prove that $\angle B + \angle F + \angle D = \angle E + \angle G$.

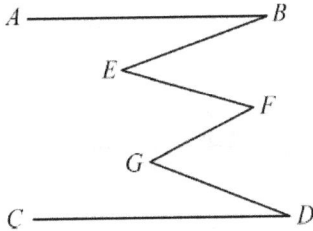

Fig. 12.9

3. As shown in Figure 12.10, $\angle ABC + \angle BCD + \angle EDC = 360°$. Prove that $AB//ED$.

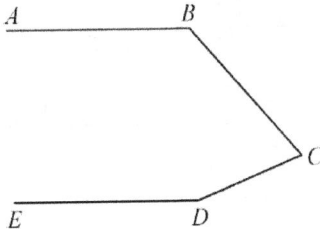

Fig. 12.10

4. As shown in Figure 12.11, $AB//CD$ and $\angle A = \angle C$. Prove that
   (1) $AD//BC$;
   (2) $\angle B = \angle D$.

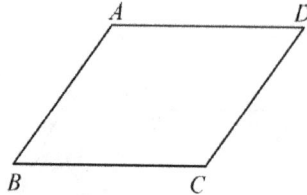

Fig. 12.11

5. As shown in Figure 12.12, $AB//CD$, $\angle 1 = \angle 2$, and $\angle EFD = 70°$. Find the degree of $\angle D$.

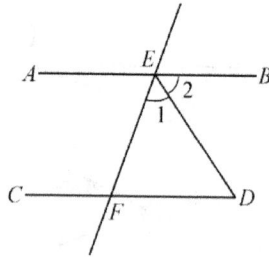

Fig. 12.12

6. As shown in Figure 12.13, given that $AB//CD//EF$, $PS \perp GH$ and intersect $GH$ at $P$. Find the degree of $\angle PSQ$ when $\angle FRG = 110°$.

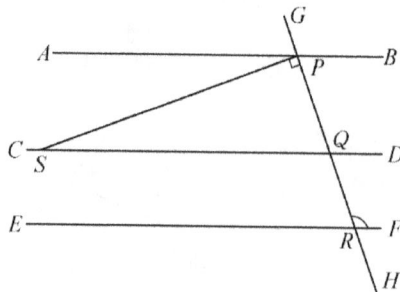

Fig. 12.13

7. As shown in Figure 12.14, given that $AB//CD$, $CD//EF$, $\angle A = 105°$, and $\angle ACE = 51°$, find the degree of $\angle E$.

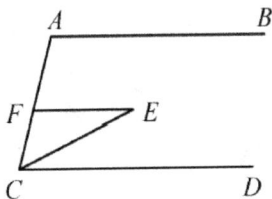

Fig. 12.14

8. As shown in Figure 12.15, in $\triangle ABC$, the bisectors of $\angle ABC$ and $\angle ACB$ intersect at $O$. Construct $EF//BC$ through $O$ that intersects $AB$ and $AC$ at $E$ and $F$, respectively. If $\angle BOC = 130°$ and $\angle ABC : \angle ACB = 3 : 2$, find the degrees of $\angle AEF$ and $\angle EFC$.

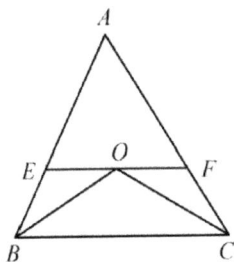

Fig. 12.15

9. As shown in Figure 12.16, $\angle ABE + \angle DEB = 180°$ and $\angle 1 = \angle 2$. Prove that $\angle F = \angle G$.

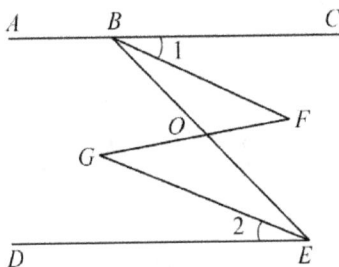

Fig. 12.16

10. As shown in Figure 12.17, $DA \perp AB$, $DE$ bisects $\angle ADC$, $CE$ bisects $\angle BCD$, and $\angle 1 + \angle 2 = 90°$. Prove that $BC \perp AB$.

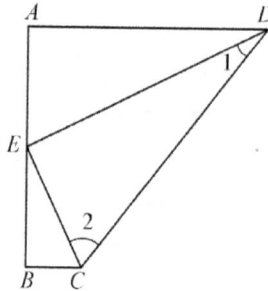

Fig. 12.17

11. There are eight straight lines on the plane where any two lines intersect. Prove that at least one of the intersecting angles is less than $23°$.

# Chapter 13

# Assigning Variables Unnecessary for Solution

Previously, we learned to solve application problems by setting up (a system of) equations. Sometimes, word problems involve a lot of quantities, and the relationship between the quantities is not very obvious. It is not easy to establish the equations directly from the unknown variables indicated by the problem. We need to assign some auxiliary variables to express those inconspicuous relationships. During the solving process, these auxiliary variables can be eliminated according to the characteristics of (the system of) equations, and then the solutions of the original problems can be obtained. This method is called assigning auxiliary variables.

**Example 1.** A pool has a drain pipe open at the bottom and several water inlet pipes of the same thickness installed at the top. When 4 water inlet pipes are open, it takes 5 h to fill the pool. When 2 water inlet pipes are open, it takes 15 h to fill the pool. Now, what is the minimum number of open inlet pipes required to fill the pool in 2 h?

**Solution** Suppose the water inflow of each inlet pipe for 1 h is $a$ and the water outflow of the drain pipe for 1 h is $b$. Suppose that $x$ open inlet pipes are needed to fill the pool in 2 h, then considering the volume of the pool,

$$\begin{cases} (4a - b) \times 5 = (2a - b) \times 15, & (13.1) \\ (xa - b) \times 2 = (2a - b) \times 15. & (13.2) \end{cases}$$

There are only two equations in this system of equations, but there are three unknowns. In general, the values of these three unknowns cannot be determined. However, due to the characteristics of this system of equations, we can find the value of $x$ without specifically finding the values of the auxiliary variables $a$ and $b$.

From (13.1), we have $4a - b = 6a - 3b$, that is,

$$a = b. \tag{13.3}$$

Substituting (13.3) into (13.2), we get

$$2(ax - a) = 15(2a - a),$$

$$2ax = 17a.$$

So, $x = 8.5$.

Of course, there cannot be half a water pipe. So, at least 9 water inlet pipes must be open in order to fill the pool within 2 h.

**Answer**    At least 9 inlet pipes must be open.

Remark: If only one unknown, $x$, which is the number of open pipes required, is assigned in this problem, then the equivalence relationship will not be obvious. Therefore, we also introduce the auxiliary variables $a$ and $b$, and set up a system of two equations with three unknowns. Then, according to the characteristics of this system of equations, we can find the value of $x$ without solving for $a$ and $b$.

**Example 2.** Bob climbs a mountain. After reaching the top of the mountain, he goes down along the original road and reaches the starting point. If Bob's uphill and downhill speeds are 3 and 5 km/h, respectively, find Bob's average speed during the whole trip.

**Solution**    Suppose that the one-way distance of the trip is $s$ km. Then, the time he spent going up the mountain is $\frac{s}{3}$ h. The time he spent going down the mountain is $\frac{s}{5}$ h. Therefore, in the whole trip, the total distance is $2s$ and the total time is $\frac{s}{3} + \frac{s}{5}$. According to the formula for average speed, we have

$$\text{average speed} = \frac{\text{total distance}}{\text{total time}} = \frac{2s}{\frac{s}{3} + \frac{s}{5}} = \frac{2s}{(\frac{1}{3} + \frac{1}{5})s}$$

$$= \frac{2}{\frac{1}{3} + \frac{1}{5}} = \frac{15}{4} (\text{ km/h}).$$

**Answer**    The average speed of Bob in the whole trip is $\frac{15}{4}$ km/h.

Remark: Do not carelessly assume that the average speed is $\frac{3+5}{2} = 4$ (km/h). This kind of mistake is due to a misunderstanding of the average speed. We know that average speed = total distance $\div$ total time, and the total distance is not given in this problem, so we introduce an auxiliary variable $s$ as the distance. After establishing the formula, we find that the value of $s$ is needed, but $s$ is naturally eliminated during the computation.

**Example 3.** Bob and Danny set off at the same time from $A$ toward $B$. (Suppose their speeds are constants.) When Bob walked halfway, Danny had only walked 16 km. When Danny walked halfway, Bob had walked 25 km. When Bob walked the whole way, how many kilometers did Danny still need to walk?

**Solution**  Suppose that the half-way length is $s$ km, Bob's speed is $x$ km/h, and Danny's speed is $y$ km/h. As the distance traveled in the same time is proportional to the speed, we have

$$\begin{cases} \dfrac{s}{16} = \dfrac{x}{y}, & (13.4) \\[2mm] \dfrac{25}{s} = \dfrac{x}{y}. & (13.5) \end{cases}$$

By (13.4) and (13.5), we find $\frac{s}{16} = \frac{25}{s}$.
That is,

$$m = s^2 = 16 \times 25 = 4^2 \times 5^2 = 20^2.$$

As $s > 0$, we have $s = 20$. When Bob finishes the trip, the difference between the locations of the two persons is

$$2 \times (20 - 16) = 8.$$

**Answer**  Danny still needs to walk 8 km.

**Example 4.** Five people, $A$, $B$, $C$, $D$, and $E$, do a project. If $A$, $B$, $C$, and $D$ work together, it can be completed in 8 days. If $B$, $C$, $D$, and $E$ work together, it can be completed in 6 days. If $A$ and $E$ work together, it can be completed in 12 days. If $A$ works alone, how many days will it take to complete the project? (Suppose that the daily workload of each person is constant.)

**Solution**  Suppose that the total workload is 1 and the daily workload that each person $A$, $B$, $C$, $D$, and $E$ can complete is $a$, $b$, $c$, $d$, and $e$, respectively. According to the problem, we get

$$\begin{cases} a + b + c + d = \dfrac{1}{8}, & (13.6) \\[2mm] b + c + d + e = \dfrac{1}{6}, & (13.7) \\[2mm] a + e = \dfrac{1}{12}. & (13.8) \end{cases}$$

By (13.6) − (13.7), we find

$$a - e = -\frac{1}{24}. \qquad (13.9)$$

By (13.8) + (13.9), we have

$$2a = \frac{1}{24},$$

$$a = \frac{1}{48}.$$

So, if $A$ works alone, then it will take $\frac{1}{a} = 48$ days to complete the work.

**Answer**   If $A$ works alone, then it will take 48 days to complete the work.

**Example 5.** As shown in Figure 13.1, there are two triangles with areas of 10 and 12 in a trapezoid. Given that the length of the upper base of the trapezoid is $\frac{2}{3}$ of the length of the lower base, find the area of the shaded part in the figure.

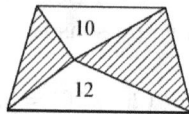

Fig. 13.1

**Solution**   Suppose that the length of the upper base of this trapezoid is $2a$ and the length of the lower base is $3a$. The height of the trapezoid is the sum of the heights of the two triangles with areas 10 and 12, that is,

$$\frac{2 \times 10}{2a} + \frac{2 \times 12}{3a} = \frac{18}{a}.$$

The area of the trapezoid is

$$\frac{1}{2}(2a + 3a) \times \frac{18}{a} = 45.$$

So, the shaded area is $45 - 10 - 12 = 23$.

**Example 6.** A three-digit number, whose tens digit is 0, is exactly equal to 67 times the sum of its digits. After interchanging the ones and hundreds digits, a new three-digit number is obtained. This new number is $m$ times the sum of its digits. Find the value of $m$.

**Solution**   Suppose that the hundreds digit of the original three-digit number is $x$ and the ones digit is $y$. According to the problem, we find

$$\begin{cases} 100x + y = 67(x+y), & (13.10) \\ 100y + x = m(x+y). & (13.11) \end{cases}$$

By (13.10) + (13.11), we get $101(x+y) = (67+m)(x+y)$,
As $x + y > 0$, we have $67 + m = 101$, $m = 34$.

**Answer**   The value of $m$ is 34.

**Example 7.** Given that $a_1, a_2, \ldots, a_{1999}, a_{2000}$ are positive numbers,

$$M = (a_1 + a_2 + \cdots + a_{1999})(a_2 + a_3 + \cdots + a_{2000}),$$

$$N = (a_1 + a_2 + \cdots + a_{2000})(a_2 + a_3 + \cdots + a_{1999}).$$

Try to compare the sizes of $M$ and $N$.

**Analysis**   At first glance, this problem does not require any auxiliary variables. But after some careful observation, we will find that in the expressions of $M$ and $N$, $a_2, \ldots, a_{1999}$ are always combined as $a_2+a_3+\cdots+a_{1999}$. If we expanded all terms by multiplying, there would be too many terms, and it would be difficult to compare the sizes. Contrarily, if we define $a_2 + a_3 + \cdots + a_{1999} = x$, the problem can be easily solved.

**Solution**   Set $x = a_2 + a_3 + \cdots + a_{1999}$, then

$$M = (a_1 + x)(x + a_{2000}) = x^2 + (a_1 + a_{2000})x + a_1 a_{2000},$$

$$N = (a_1 + x + a_{2000})x = x^2 + (a_1 + a_{2000})x.$$

So, $M - N = a_1 a_{2000}$.
As $a_1 > 0$ and $a_{2000} > 0$, we have $a_1 a_{2000} > 0$, that is, $M - N > 0$, or $M > N$.

**Answer**   $M > N$.

Remark:   In summary, sometimes the equivalence relations are difficult to find or solve directly, and it becomes much easier by introducing some auxiliary variables. During the process of computation, we can manage to eliminate the auxiliary variables and find the answers.

## Exercises

1. Alice goes up a mountain at 4 km/h. After reaching the top of the mountain, she immediately goes down the mountain along the same road and walks at 6 km/h until she returns to the starting point. Find Alice's average speed during the whole trip.

2. A ship travels 10 km/h when sailing downstream and 6 km/h when sailing upstream. What is the average speed of the ship during the round trip?

3. A ship was sailing upstream in a river. The crew accidentally dropped a wooden barrel into the water. They found that the barrel was missing after 5 min and turned around to chase it. How long did it take to catch up with the wooden barrel?

4. Cup $A$ is filled with water, and Cup $B$ is filled with alcohol. There is as much water as alcohol. First, pour some water from Cup $A$ into Cup $B$. Then, pour the same amount of mixed liquid from Cup $B$ into Cup $A$. Then, pour some mixed liquid from Cup $A$ into Cup $B$, then put the same amount from Cup $B$ into Cup $A$, and go on like this for 100 times. In the end, is there more alcohol in Cup $A$ or more water in Cup $B$?

5. Consider the polynomial $ax^5 + bx^3 + cx + 3$. When $x = 3$, its value is 48. Find the value of this polynomial when $x = -3$.

6. Four people $A$, $B$, $C$, and $D$ do a project. If $A$, $B$, and $C$ work together, then it can be completed in 10 days. If $A$, $C$, and $D$ work together, then it can be completed in 8 days. If $B$ and $D$ work together, then it can be completed in 20 days. If $D$ works alone, how many days will it take to complete the project?

7. Stations $A$, $B$, and $C$ are on a straight line. Station $B$ is exactly at the midpoint of stations $A$ and $C$; that is, the distance from $B$ to $A$ is equal to that from $B$ to $C$. Alice and Bob travel from $A$ and $C$, respectively. They start at the same time and walk toward each other. Alice walks 100 m after passing station $B$ and meets Bob for the first time. Then, they continue to move forward. Alice returns immediately after reaching station $C$, and after passing station $B$ for 300 m, she catches up with Bob. What is the distance between stations $A$ and $B$?

8. As shown in Figure 13.2, a rectangle is divided into four sub-rectangles by two straight lines. Three of these have areas of 20, 25, and 30. What is the area of the other (shaded part in the figure) rectangle?

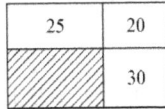

Fig. 13.2

9. There are three kinds of goods $A$, $B$, and $C$. If you buy 3 pieces of $A$, 7 pieces of $B$, and 1 piece of $C$, then the total cost is 300 yuan. If you buy 4 pieces of $A$, 10 pieces of $B$, and 1 piece of $C$, then the total cost is 400 yuan. If you buy 1 piece of each of $A$, $B$, and $C$, how much is the total cost?

10. Two cups, $A$ and $B$, are respectively filled with 16 and 24 g of mixed solutions of alcohol and water with different concentrations. Pour out the mixed solutions $A_1$ and $B_1$ of equal weight from the two cups of $A$ and $B$, respectively. Pour $A_1$ into cup $B$ and pour $B_1$ into cup $A$. Now, the concentrations of alcohol in cups $A$ and $B$ are equal. How many grams does $A_1$ or $B_1$ weigh?

11. A person started breakfast after 8 o'clock, and he found that the angle between the minute hand and the hour hand on the clock was 25°. After he finished his breakfast, he found that the time on the clock was still before 9 o'clock, and the angle between the two hands was still 25°. How long did he spend eating breakfast?

12. There is a reservoir with daily water flowing in and out at the same time. The water in the reservoir will run out in 40 days. Recent rainfall in the reservoir area then increased the daily amount of water flowing into the reservoir by 20%. If the daily amount of water flowing out of the reservoir increases by 10%, the water will again run out in 40 days. If the daily amount of water flowing out of the reservoir does not change, in how many days will the water run out?

13. Each of three mixtures consists of some of the three ingredients $A$, $B$, and $C$. The first mixture contains only ingredients $A$ and $B$ in a weight

ratio of 3 : 5. The second mixture contains only ingredients $B$ and $C$ in a weight ratio of 1 : 2. The third mixture contains only ingredients $A$ and $C$ in the weight ratio of 2 : 3. What is the ratio to combine these mixtures so that the weight ratio of the three ingredients $A$, $B$, and $C$ in the resulting mixture is 3 : 5 : 2?

14. The age of Bob's grandfather is a two-digit number. The number obtained by interchanging the two digits is exactly the age of Bob's father. It is also known that the difference between their ages is four times Bob's age. Find Bob's age.

# Chapter 14

# Undetermined Coefficients

If two polynomials in $x$,

$$a_n x^n + a_{n-1} x^{n-1} + \cdots + a_0$$

and

$$b_n x^n + b_{n-1} x^{n-1} + \cdots + b_0,$$

are exactly the same, then the corresponding coefficients are all equal:

$$a_n = b_n, a_{n-1} = b_{n-1}, \ldots, a_0 = b_0.$$

If we replace the variable $x$ with any value, the values of the two polynomials are equal; that is, the two polynomials are identical.

Conversely, if two polynomials are identical, that is, if the variable $x$ is replaced by any value, the values of the two polynomials are equal, then they have the same degree and the corresponding coefficients of the two polynomials are equal. This is called the identity theorem for polynomials.

When two polynomials are identical, the corresponding coefficients are equal. If one of the polynomials has some undetermined coefficients, these undetermined coefficients can be obtained from the other polynomial.

**Example 1.** For what values of $m$ and $n$ are the polynomials $x^2 + mx + n$ and $(x - 2)(x + 3)$ identical?

**Solution** By the given condition (use the symbol "$\equiv$" to indicate the identity),

$$x^2 + mx + n \equiv (x - 2)(x + 3) \equiv x^2 + x - 6.$$

After comparing the corresponding coefficients, we have

$$m = 1, \quad n = -6.$$

**Another Solution**    Replacing $x$ in the polynomials with 0 and 1, respectively, we have

$$\begin{cases} n = -6, \\ 1 + m + n = -4. \end{cases}$$

After solving it, we have $m = 1$, $n = -6$.

Remark:    According to the situation, either or both of the above two methods can be used. The values of $x$ in the second solution can be selected arbitrarily. Of course, we wish that the generated (system of) equations were as simple as possible. In this problem, $x$ can also be assigned as $-1$ and 2.

**Example 2.** Given the polynomial

$$x^4 + x^3 + x^2 + 2 \equiv (x^2 + mx + 1)(x^2 + nx + 2),$$

find the values of $m$ and $n$.

**Solution**

$$x^4 + x^3 + x^2 + 2$$
$$\equiv (x^2 + mx + 1)(x^2 + nx + 2)$$
$$\equiv x^4 + (m + n)x^3 + (mn + 3)x^2 + (2m + n)x + 2.$$

After comparing the corresponding coefficients, we have

$$\begin{cases} m + n = 1, & (14.1) \\ mn + 3 = 1, & (14.2) \\ 2m + n = 0. & (14.3) \end{cases}$$

On $(14.3) - (14.1)$, we find

$$m = -1.$$

Substitute $m = -1$ into $(14.1)$ to get

$$n = 2.$$

When $m = -1$ and $n = 2$, $(14.2)$ obviously holds.
So, $m = -1$ and $n = 2$.

Remark:    Two first-degree equations $(14.1)$ and $(14.3)$ are enough to find the values of $m$ and $n$. The third equation 14.2 is used for checking answers.

**Example 3.** Suppose that the product of $ax^2 + bx + 1$ and $2x^2 - 3x + 1$ does not contain a cubic term or a first-order term. Try to find the values of $a$ and $b$.

**Solution** The cubic term of the product is $ax^2 \cdot (-3x) + bx \cdot (2x^2) = (2b - 3a)x^3$. Similarly, the first-order term is $bx \cdot 1 + 1 \cdot (-3x) = (b - 3)x$. It is given that the coefficients of these two terms are both zero. Therefore, we get

$$\begin{cases} 2b - 3a = 0, \\ b - 3 = 0. \end{cases}$$

After solving it, we have

$$\begin{cases} a = 2, \\ b = 3. \end{cases}$$

Remark: In order to get the expression of a certain power of $x$ in the product, it is not necessary to write out the whole product. We can just write out all the terms with $x$ raised to that power.

**Example 4.** Given that

$$6x^2 - 7xy - 3y^2 + 14x + y + a \equiv (2x - 3y + b)(3x + y + c),$$

try to find the values of $a$, $b$, and $c$.

**Solution** By the meaning of this problem, we have

$$6x^2 - 7xy - 3y^2 + 14x + y + a$$
$$\equiv (2x - 3y + b)(3x + y + c)$$
$$\equiv 6x^2 - 7xy - 3y^2 + (3b + 2c)x + (b - 3c)y + bc.$$

After comparing the corresponding coefficients, we find

$$\begin{cases} 3b + 2c = 14, & (14.4) \\ b - 3c = 1, & (14.5) \\ bc = a. & (14.6) \end{cases}$$

By (14.4) and (14.5), we get

$$b = 4, c = 1.$$

Substitute $b = 4$ and $c = 1$ into (14.6) to find

$$a = 4.$$

So, $a = 4$, $b = 4$, and $c = 1$.

**Remark:** The polynomial given in this example is a second-degree polynomial with two variables. Similar to polynomials with one variable, if two polynomials with two variables are identical, then their corresponding coefficients are all equal. Therefore, we can compare the corresponding coefficients to find the undetermined coefficients.

**Example 5.** If $5x^2 - kx + 7$ is divided by $5x - 2$ and the remainder is 6, find the value of $k$ and the quotient.

**Solution**   As $5x^2 \div 5x = x$, the quotient's highest degree term is a first-order term and the coefficient is 1. Suppose the quotient is $x + m$, and we find

$$5x^2 - kx + 7 \equiv (5x - 2)(x + m) + 6, \tag{14.7}$$

that is,

$$5x^2 - kx + 7 \equiv 5x^2 + (5m - 2)x - 2m + 6, \tag{14.8}$$

After comparing the corresponding coefficients, we have

$$\begin{cases} -k = 5m - 2, & (14.9) \\ 7 = -2m + 6. & (14.10) \end{cases}$$

After solving it, we get

$$\begin{cases} m = -\frac{1}{2}, \\ k = \frac{9}{2}. \end{cases}$$

So, $k = \frac{9}{2}$, and the quotient is $x - \frac{1}{2}$.

**Another Solution**   On (14.7), we let $x = \frac{2}{5}$ to get

$$\frac{4}{5} - \frac{2k}{5} + 7 = 6,$$

and we find $k = \frac{9}{2}$.

Let $x = 0$ in (14.7), we get $7 = -2m + 6$, and $m = -\frac{1}{2}$.

**Example 6.** Express $5x^3 - 6x^2 + 10$ in the form $a(x - 1)^3 + b(x - 1)^2 + c(x - 1) + d$.

**Solution**   Clearly, we have

$$5x^3 - 6x^2 + 10$$

$$\equiv a(x-1)^3 + b(x-1)^2 + c(x-1) + d$$

$$\equiv ax^3 + (b-3a)x^2 + (3a-2b+c)x + (b+d-a-c).$$

After comparing the corresponding coefficients, we get

$$\begin{cases} a = 5, \\ b - 3a = -6, \\ 3a - 2b + c = 0, \\ b + d - a - c = 10. \end{cases}$$

After solving it, we find

$$a = 5, \quad b = 9, \quad c = 3, \quad d = 9.$$

So, $5x^3 - 6x^2 + 10 = 5(x-1)^3 + 9(x-1)^2 + 3(x-1) + 9$.

Remark:   In addition to the method of undetermined coefficients, we can also use the method of variable substitution as follows:

Let $y = x - 1$, then $x = y + 1$, and substitute it in the original formula to obtain

$$5x^3 - 6x^2 + 10$$

$$= 5(y+1)^3 - 6(y+1)^2 + 10$$

$$= 5y^3 + 9y^2 + 3y + 9$$

$$= 5(x-1)^3 + 9(x-1)^2 + 3(x-1) + 9.$$

This method is simpler than solving a system of equations.

**Example 7.** There is a cubic polynomial with one variable, $x$. When $x$ takes the values 0, 1, $-1$, and 2, the values of this polynomial are 2, 2, 0, and 6, respectively. Find this polynomial.

**Solution**   Suppose the polynomial is $ax^3 + bx^2 + cx + d$. Then, we have

$$\begin{cases} d = 2, \\ a + b + c + d = 2, \\ -a + b - c + d = 0, \\ 8a + 4b + 2c + d = 6. \end{cases}$$

On solving it, we get

$$a = 1, \quad b = -1, \quad c = 0, \quad d = 2.$$

So, the polynomial is $x^3 - x^2 + 2$.

## Exercises

1. What are the values of $p$ and $q$ so that the polynomials $x^2 + px + q$ and $(x+2)(x-5)$ are identical?

2. Given a polynomial $ax + b$, when $x$ takes the values of 1 and $-1$, the values of this polynomial are 11 and $-3$, respectively. Find the value of this polynomial when $x = 3$.

3. Given a quadratic polynomial with three terms, $x^2 + px + q$, when $x$ takes values of 0 and 1, its values are 2 and 9, respectively. Find this quadratic polynomial.

4. A quadratic polynomial consists of three terms involving $x$. When $x$ takes values of 0, 1, and $-1$, the values of this polynomial are 1, 4, and 2, respectively. Find this quadratic polynomial.

5. Given $x^3 + ax^2 + bx + c \equiv (x-1)^2(x+1)$. Find the values of $a$, $b$, and $c$.

6. If $x^4 - 6x^3 + 13x^2 - 12x + k \equiv (x^2 + mx + n)^2$, then find the values of $m$, $n$, and $k$.

7. Suppose that $6x^2 - 5xy - 4y^2 - 11x + 22y + m \equiv (2x+y+k)(3x-4y+l)$. Find the values of $m$, $k$, and $l$.

8. Express $3x^2 - 4x + 7$ in the form of $a(x+1)^2 + b(x+1) + c$.

9. Suppose $\frac{x+4}{x^2+5x+6} \equiv \frac{A}{x+2} + \frac{B}{x+3}$. Try to find the values of $A$ and $B$.

10. Suppose that the product of $3x^2 + ax + b$ and $2x^2 - 4x + 5$ does not contain the cubic term and the first-order term. Find the values of $a$ and $b$.

11. Suppose that $\frac{x^2}{x(x-1)(x+1)} \equiv \frac{A}{x} + \frac{B}{x-1} + \frac{C}{x+1}$. Find the values of $A$, $B$, and $C$.

12. Suppose that $\frac{x^2-2x+5}{(x-2)^2(x^2+1)} \equiv \frac{A}{x-2} + \frac{B}{(x-2)^2} + \frac{Cx+D}{x^2+1}$. Find the values of $A$, $B$, $C$, and $D$.

13. Given that
$$(x^2 + ax + b)^{10} \equiv (x+5)^{20} - (cx+d)^{20}.$$
Find the values of $a$, $b$, $c$, and $d$.

14. Given that the polynomial $ax^3 + bx^2 + cx + d$ is divisible by $x^2 + p$, prove that $ad = bc$.

15. Suppose that the coefficients of $x^3 + bx^2 + cx + d$ are all integers. If $bd + cd$ is an odd number, prove that this polynomial cannot be expressed as a product of two polynomials with integer coefficients.

# Chapter 15

# Synthetic Division and Polynomial Remainder Theorem

Synthetic division and the polynomial remainder theorem are common tools for studying polynomial divisions.

**Example 1.** Find the quotient and the remainder of the polynomial $3x^2 + 5x - 7$ divided by $x + 2$.

**Solution**   We first use the general long division to compute

$$
\begin{array}{r}
3x - 1 \phantom{000} \\
x + 2 \overline{\smash{)}3x^2 + 5x - 7} \\
\underline{3x^2 + 6x} \phantom{0000} \\
-x - 7 \phantom{00} \\
\underline{-x - 2} \phantom{00} \\
-5
\end{array}
$$

So, the quotient is $3x - 1$ and the remainder is $-5$.

From the calculation, we realize that the above operation is actually performed on the coefficients, and so we can omit the letters and express the above division in the following convenient way:

$$
\begin{array}{r|rrr}
-2 & 3 & +5 & -7 \\
\hline
& 3 & -1 & \multicolumn{1}{|r}{-5}
\end{array}
$$

The quotient is $3x - 1$, and the remainder is $-5$.

When the divisor is a first-degree polynomial $x - a$, this simple division method can be used, which is called the synthetic division method. The computation process is as follows:

(1) Arrange the dividend in the order of descending powers of $x$, and write the coefficients accordingly from left to right. If there is a missing term, then it must be filled with "0." For the above example, we write 3, +5, and $-7$.

(2) Write $a$ (the opposite number of $-a$) in the divisor on the left side of the coefficients of the dividend, and separate them with a vertical line. For the above example, write $-2$ on the left and put a vertical line between $-2$ and 3.

(3) Write the first coefficient of the dividend as the first number of the quotient right below it. Multiply it by $a$, and add the second coefficient of the dividend to get the second number in the second row. For the above example, it is $(-2) \cdot 3 + 5 = -1$. Then, multiply the second number by $a$, and add the third coefficient of the dividend to get the third number in the second row (which is $(-2) \cdot (-1) + (-7) = -5$) ..., and so on. The last number we get is the remainder, which is right below the last coefficient of the dividend (which is $-5$). We use a vertical line to separate it from the other coefficients on the left, which are for the quotient (here, the numbers 3 and $-1$ refer to $3x - 1$).

An algebraic expression of $x$ is often represented by $f(x)$ or $g(x)$; for example, the algebraic expression $2x^2 + x - 3$ is represented by $f(x)$, which can be written as

$$f(x) = 2x^2 + x - 3.$$

Here, $f(1)$ means the value of the algebraic formula $2x^2 + x - 3$ when $x = 1$, that is, $f(1) = 2 \times 1^2 + 1 - 3 = 0$. Similarly, we have $f(0) = 2 \times 0^2 + 0 - 3 = -3$, $f(-1) = 2 \times (-1)^2 + (-1) - 3 = -2$, and so on.

$f(x)$ can represent any algebraic expression of $x$. But in the same problem, different algebraic expressions should be represented by different symbols, such as $f(x)$, $g(x)$, $q(x)$, and $r(x)$.

Using the above notation, in division, we have

$$f(x) = g(x) \cdot q(x) + r(x). \tag{15.1}$$

Here, $f(x)$ represents the dividend, $g(x)$ represents the divisor, $q(x)$ represents the quotient, and $r(x)$ represents the remainder. The degree of the remainder $r(x)$ is less than that of the divisor $g(x)$.

If $g(x)$ is a first-degree formula $x - a$, then the degree of $r(x)$ is less than 1, so $r(x)$ can only be a constant (0 or a non-zero constant). Now, the remainder formula is also called the remainder, denoted by $r$, that is,

$$f(x) = (x - a) \cdot q(x) + r. \qquad (15.2)$$

Let $x = a$ in (15.2). We find

$$f(a) = r.$$

Therefore, we have the following important theorems.

**The Polynomial Remainder Theorem.** If a polynomial $f(x)$ is divided by $x - a$, then the remainder is equal to $f(a)$.

For example, the remainder of $f(x) = 3x^2 + 5x - 7$ divided by $x + 2$ is

$$f(-2) = 3 \times (-2)^2 + 5 \times (-2) - 7 = -5.$$

This is the same remainder we found earlier using the synthetic division method.

Also, by (15.2), if $f(x)$ is divisible by $x - a$, then $r = 0$. Conversely, if $r = 0$, then $f(x)$ is divisible by $x - a$. Therefore, we have the following theorem.

**The Polynomial Factor Theorem.** If a polynomial $f(x)$ is divisible by $x - a$, that is, $f(x)$ has a factor $x - a$, then $f(a) = 0$. Conversely, if $f(a) = 0$, then $x - a$ must be a factor of $f(x)$.

**Example 2.** Find the quotient and remainder of $2x^4 - 3x^3 - x^2 + 5x + 6$ divided by $(x + 1)$.

**Solution**   Use the synthetic division to get the following:

$$
\begin{array}{r|rrrrr}
-1 & 2 & -3 & -1 & +5 & +6 \\
   &   & 2 & -5 & 4 & 1 & 5 \\
\hline
   & 2 & -5 & 4 & 1 & \multicolumn{1}{|r}{5}
\end{array}
$$

So, the quotient is $2x^3 - 5x^2 + 4x + 1$ and the remainder is 5.

**Example 3.** Find the quotient and remainder of the polynomial $f(x) = 3x^3 + 5x^2 - 2x^4 - 5$ divided by $x - 2$.

**Solution**   We first arrange the terms by descending powers of $x$:

$$f(x) = 3x^3 + 5x^2 - 2x^4 - 5$$

$$= -2x^4 + 3x^3 + 5x^2 + 0 \cdot x - 5.$$

Use the synthetic division to obtain the following:

$$
\begin{array}{r|rrrrr}
2 & -2 & +3 & +5 & 0 & -5 \\
\hline
& -2 & -1 & 3 & 6 & \boxed{7}
\end{array}
$$

So, the quotient is $-2x^3 - x^2 + 3x + 6$ and the remainder is 7.

Remark: Note that when we first arrange $f(x)$ in descending powers, if there are missing terms, we should fill them with zeros.

**Example 4.** Use the synthetic division to compute

$$(6x^4 - 7x^3 - x^2 + 8) \div (2x + 1).$$

**Solution**    $2x + 1 = 2(x + \frac{1}{2})$. We first compute $6x^4 - 7x^3 - x^2 + 8$ divided by $x + \frac{1}{2}$:

$$
\begin{array}{r|rrrrr}
-\frac{1}{2} & 6 & -7 & -1 & 0 & +8 \\
\hline
& 6 & -10 & 4 & -2 & \boxed{9}
\end{array}
$$

So, we have

$$6x^4 - 7x^3 - x^2 + 8$$

$$= \left(x + \frac{1}{2}\right)(6x^3 - 10x^2 + 4x - 2) + 9$$

$$= 2\left(x + \frac{1}{2}\right) \cdot \frac{1}{2}(6x^3 - 10x^2 + 4x - 2) + 9$$

$$= (2x + 1)(3x^3 - 5x^2 + 2x - 1) + 9.$$

Thus, the quotient is $3x^3 - 5x^2 + 2x - 1$ and the remainder is 9.

Remark:    If the divisor is a first-degree polynomial, but the coefficient of $x$ is not 1, that is, the divisor is $ax + b$ ($a \neq 0$ and $a \neq 1$), we can first divide $f(x)$ by $x + \frac{b}{a}$ (at this point, we can use the synthetic division method). We then have $f(x) = (x + \frac{b}{a}) \cdot q(x) + r$, where $r = f(-\frac{b}{a})$. Therefore, we have

$$f(x) = a\left(x + \frac{b}{a}\right) \cdot \frac{1}{a} \cdot q(x) + r$$

$$= (ax + b)\left(\frac{1}{a} \cdot q(x)\right) + r.$$

Thus, the quotient is $\frac{1}{a} \cdot q(x)$ and the remainder is still $r$.

## Example 5.

(1) Find the remainder of $f(x) = 7x^5 - 4x^4 - 6x^2 + 5$ divided by $x - 1$.
(2) Find the remainder of $f(x) = 7x^5 - 4x^4 - 6x^2 + 5$ divided by $2x - 2$.

## Solution

(1) By the remainder theorem, the remainder is

$$f(1) = 7 \times 1^5 - 4 \times 1^4 - 6 \times 1^2 + 5 = 2.$$

(2) From the remark in Example 4, the remainder of $f(x)$ divided by $ax+b$ is the same as the remainder of $f(x)$ divided by $x + \frac{b}{a}$. So, the remainder of $7x^5 - 4x^4 - 6x^2 + 5$ divided by $2x - 2$ is the same as that of $7x^5 - 4x^4 - 6x^2 + 5$ divided by $x - 1$. Thus, the remainder is

$$f(1) = 7 \times 1^5 - 4 \times 1^4 - 6 \times 1^2 + 5 = 2.$$

**Remark:** This problem can also be solved using the synthetic division method.

**Example 6.** The remainders of the polynomial $f(x)$ divided by $x - 1$ and $x - 2$ are 3 and 5, respectively. Find the remainder of $f(x)$ divided by $(x - 1)(x - 2)$.

**Solution** From the given problem statement and the polynomial remainder theorem, we find

$$f(1) = 3, f(2) = 5.$$

Suppose that the quotient of $f(x)$ divided by $(x - 1)(x - 2)$ is $q(x)$ and the remainder is $cx + d$ (as the divisor's degree is two, so the remainder's degree is at most one), then we get

$$f(x) = (x - 1)(x - 2) \cdot q(x) + (cx + d).$$

So, we have

$$\begin{cases} f(1) = c + d = 3, \\ f(2) = 2c + d = 5. \end{cases}$$

$$(15.3)$$
$$(15.4)$$

By (15.3) and (15.4), we find

$$\begin{cases} c = 2, \\ d = 1. \end{cases}$$

Thus, the remainder is $2x + 1$.

**Remark:** The coefficients $c$ and $d$ of the remainder $cx + d$ of this problem need to be determined. We can use the polynomial remainder theorem to find $c$ and $d$.

Generally, for unequal constants $a$ and $b$, suppose the quotient of $f(x)$ divided by $(x - a)(x - b)$ is $q(x)$ and the remainder is $cx + d$. Then, we have

$$f(x) = (x - a)(x - b)q(x) + (cx + d). \qquad (15.5)$$

So,

$$\begin{cases} f(a) = ca + d, \\ f(b) = cb + d, \end{cases}$$

On solving it, we find

$$\begin{cases} c = \dfrac{f(a) - f(b)}{a - b}, & (15.6) \\[2mm] d = \dfrac{af(b) - bf(a)}{a - b}; & (15.7) \end{cases}$$

that is, the remainder is

$$\frac{f(a) - f(b)}{a - b}x + \frac{af(b) - bf(a)}{a - b}. \qquad (15.8)$$

**Example 7.** Let $a$ and $b$ be unequal constants. If a polynomial $f(x)$ is divisible by $x - a$ and $x - b$, then prove that $f(x)$ is also divisible by $(x - a)(x - b)$.

**Proof** As $f(x)$ is divisible by $x - a$ and $x - b$, so $f(a) = 0$ and $f(b) = 0$.

Then, in the above (15.6) and (15.7), both $c$ and $d$ are zero, and the remainder in (15.5) is 0; that is, $f(x)$ is divisible by $(x - a)(x - b)$.

**Remark:** The conclusion of this problem is very useful.

**Example 8.** Try to find the values of $a$ and $b$ such that $f(x) = 2x^4 - 3x^3 + ax^2 + 5x + b$ is divisible by $(x + 1)(x - 2)$.

**Solution**   As $f(x)$ is divisible by $(x+1)(x-2)$, $f(x)$ is divisible by $x+1$ and $x-2$. Using the factor theorem, we find

$$f(-1) = 2 \times (-1)^4 - 3 \times (-1)^3 + a \times (-1)^2 + 5 \times (-1) + b$$

$$= a + b = 0,$$

$$f(2) = 2 \times 2^4 - 3 \times 2^3 + a \times 2^2 + 5 \times 2 + b$$

$$= 4a + b + 18 = 0;$$

that is,

$$\begin{cases} a + b = 0, \\ 4a + b + 18 = 0. \end{cases}$$

On solving it, we have $a = -6$ and $b = 6$.

**Example 9.** Prove that $a^3 + b^3 + c^3 - 3abc$ has a factor of $a + b + c$.

**Analysis**   Treat $a^3 + b^3 + c^3 - 3abc$ as a polynomial of $a$, and arrange it in descending powers as $a^3 + 0 \cdot a^2 - 3bca + (b^3 + c^3)$. The problem requires us to prove that the first-degree polynomial $a + (b + c)$ of $a$ divides this third-degree polynomial.

**Proof**   Use the synthetic division method:

$$
\begin{array}{r|cccc}
-(b+c) & 1 & 0 & -3bc & b^3 + c^3 \\
& & -(b+c) & b^2 - bc + c^2 & \\
\hline
& 1 & -(b+c) & b^2 - bc + c^2 & 0
\end{array}
$$

As the remainder of $a^3 + b^3 + c^3 - 3abc$ divided by $a + b + c$ is $0$, $a^3 + b^3 + c^3 - 3abc$ has a factor of $a + b + c$. From the above division, we have

$$a^3 + b^3 + c^3 - 3abc = (a + b + c)(a^2 + b^2 + c^2 - ab - bc - ca). \quad (15.9)$$

Remark:   Equation (15.9) is a multiplication formula.

## Exercises

1. Compute the quotient and remainder of $3x^3 - 5x + 6$ divided by $(x - 2)$.
2. Find the quotient and remainder of $2x^3 + 5x^2 - 4x^4 + 8$ divided by $x + 3$.
3. Use the synthetic division to compute

$$(-6x^4 - 7x^2 + 8x + 9) \div (2x - 1).$$

4. Use the synthetic division to compute

$$(27x^3 - 9x^2 + 5x - 2) \div (3x - 2).$$

5. Find the remainder of $6x^5 - 4x^3 + 5x^2 + 3x + 8$ divided by $x + 1$.

6. Suppose that $f(x) = x^4 + 3x^3 + 8x^2 - kx + 11$ is divisible by $x + 3$. Try to find the value of $k$.

7. Suppose that $f(x) = 2x^3 + x^2 + kx - 2$ is divisible by $2x + \frac{1}{2}$. Find the value of $k$.

8. Suppose that $f(x) = 3x^5 - 17x^4 + 12x^3 + 6x^2 + 9x + 8$. Find the value of $f(-\frac{1}{3})$.

9. Suppose that $f(x) = x^4 - ax^2 - bx + 2$ is divisible by $(x + 1)(x - 2)$. Find the values of $a$ and $b$.

10. Find the remainder of $f(x) = 3x^4 - 8x^3 + 5x^5 - x + 8$ divided by $2x - 4$.

11. If the remainder of $f(x) = 2x^3 - 3x^2 + ax + b$ divided by $x + 1$ is 7 and the remainder of $f(x)$ divided by $x - 1$ is 5, then find the values of $a$ and $b$.

12. Suppose that $f(x) = x^2 + mx + n$ (where $m$ and $n$ are both integers) is not only a factor of the polynomial $x^4 + 6x^2 + 25$ but also a factor of the polynomial $3x^4 + 4x^2 + 28x + 5$. Find $f(x)$.

13. The remainders of $f(x)$ divided by $(x - 1)$, $(x - 2)$, and $(x - 3)$ are 1, 2, and 3, respectively. Find the remainder of $f(x)$ divided by $(x - 1)(x - 2)(x - 3)$.

14. Given that the polynomial $f(x) = ax^3 + bx^2 - 8x - 12$ is divisible by $x - 2$ and $x - 3$, try to find the values of $a$ and $b$, and also find the quotient of $f(x)$ divided by $(x - 2)(x - 3)$.

15. If $x^5 - 5qx + 4r$ is divisible by $(x - 2)^2$, then find the values of $q$ and $r$.

16. When a third-degree polynomial $f(x)$ is divided by $x^2 - 1$, the remainder is $2x - 5$, and when divided by $x^2 - 4$, the remainder is $-3x + 4$. Find this third-degree polynomial.

17. There is a third-degree polynomial $f(x)$ with integer coefficients, and there are three distinct integers $a_1$, $a_2$, and $a_3$ such that

$$f(a_1) = f(a_2) = f(a_3) = 1.$$

Also, suppose $b$ is an integer not equal to $a_1$, $a_2$, or $a_3$. Prove that $f(b) \neq 1$.

# Chapter 16

# Simplifying and Evaluating an Algebraic Formula

A formula formed by connecting numbers and letters representing numbers with basic operation symbols is called an algebraic formula. When the letters in the formula are replaced by numerical values, the result computed according to the algorithm given by the algebraic formula is called the value of the algebraic formula. Therefore, the value of an algebraic formula is determined by the values of the letters contained in it and changes with the values of the letters. It is worth noting that when we choose the values of the letters in the algebraic formula, the algebraic formula must be meaningful.

The value of the algebraic formula can be obtained by directly substituting the value of the letter into it and computing. However, for very complicated algebraic formulas, it is often necessary to simplify and then evaluate. Sometimes, mathematical methods such as algebraic transformation, elimination, and parameter setting are used.

**Example 1.** Given that $x$ is the largest negative integer and $y$ is the rational number that has the smallest absolute value, find the value of the algebraic formula $3x^3 - 10x^2y + 5xy^2 - 13y^3$.

**Solution**  As $x$ is the largest negative integer, $x = -1$. As $y$ is the rational number that has the smallest absolute value, $y = 0$. Therefore,

$$3x^3 - 10x^2y + 5xy^2 - 13y^3$$

$$= 3 \times (-1)^3 - 10 \times (-1)^2 \times 0 + 5 \times (-1) \times 0^2 - 13 \times 0^3$$

$$= -3.$$

So, the value is $-3$.

Remark: For a relatively simple evaluation of algebraic formulas, as long as the value of the letter is substituted for the calculation, the problem can be solved. Of course, sometimes you need to know some common knowledge, such as, in this example, the largest negative integer and the rational number with the smallest absolute value.

**Example 2.** Given that, when $x = 5$, the value of the algebraic formula $ax^2 + bx - 5$ is 10. Find the value of $ax^2 + bx + 5$ when $x = 5$.

**Solution**   The two formulas only differ in the constant terms. So, when $x = 5$, we have

$$ax^2 + bx + 5$$
$$= (ax^2 + bx - 5) + 10$$
$$= 10 + 10$$
$$= 20.$$

Remark: Attention should be paid to observe the relationship between the two algebraic formulas. In this problem, the coefficients $a$ and $b$ are not necessary and cannot be solved.

**Example 3.** Given that $a + b = 1$, find the value of the algebraic formula $a^3 + 3ab + b^3$.

**Solution**   Use the substitution method. Since $a + b = 1$, we have $b = 1 - a$. So,

$$a^3 + 3ab + b^3$$
$$= a^3 + 3a(1 - a) + (1 - a)^3$$
$$= a^3 + 3a - 3a^2 + 1 - 3a + 3a^2 - a^3$$
$$= 1.$$

Remark: To find the value of an algebraic formula with a certain condition, we usually change the condition and evaluate by substitution.

In addition, we often transform the algebraic formula and substitute the condition into it during the process of evaluation. For example, the following are clever solutions for this problem:

$$a^3 + 3ab + b^3$$
$$= (a^3 + b^3) + 3ab$$
$$= (a + b)(a^2 - ab + b^2) + 3ab$$
$$= (a^2 - ab + b^2) + 3ab$$
$$= a^2 + 2ab + b^2$$
$$= (a + b)^2$$
$$= 1;$$

and

$$a^3 + 3ab + b^3$$
$$= a^3 + 3ab(a + b) + b^3$$
$$= (a + b)^3$$
$$= 1.$$

**Example 4.** Consider the algebraic formula $ax^3 + bx + c$. When $x = 0$, its value is 2. When $x = 3$, its value is 1. Find its value when $x = -3$.

**Solution** As $x = 0$, its value is 2. So, we have

$$a \times 0^3 + b \times 0 + c = 2,$$

that is, $c = 2$.

Note that the values of $ax^3 + bx$ are opposite numbers when $x = -3$ and $x = 3$. So, when $x = -3$, we find

$$ax^3 + bx + c = a \times (-3)^3 + b \times (-3) + 2$$
$$= -(a \times 3^3 + b \times 3 + 2) + 4$$
$$= -1 + 4$$
$$= 3.$$

Consider this: if the formula only contains even power terms of $x$ in this example, then what will happen?

**Example 5.** If $x^2 - 3x - 1 = 0$, then find the value of $2x^3 - 3x^2 - 11x + 8$.

**Solution**

$$2x^3 - 3x^2 - 11x + 8$$
$$= 2x(x^2 - 3x - 1) + 3(x^2 - 3x - 1) + 11$$
$$= 2x \times 0 + 3 \times 0 + 11$$
$$= 11.$$

Remark: When evaluating an algebraic formula, if the value of the letter is not explicitly given or is difficult to find, then it cannot be directly substituted into the calculation. In such cases, according to the characteristics of the problem, the desired algebraic formula should be properly transformed, and then the given conditions (such as the value of an algebraic formula) can be substituted as a whole to obtain a simple answer.

This problem can also be regarded as a division, that is, $2x^3 - 3x^2 - 11x + 8$ is divided by $x^2 - 3x - 1$, and the remainder is 11.

**Example 6.** Given that $a - b = 2$ and $b - c = 1$, find the value of the algebraic formula $a^2 + b^2 + c^2 - ab - bc - ca$.

**Solution**   From the given conditions, we get

$$a = b + 2, c = b - 1.$$

Substitute them into the computation to find that the original formula

$$= (b + 2)^2 + b^2 + (b - 1)^2 - (b + 2)b - b(b - 1) - (b + 2)(b - 1)$$
$$= 7.$$

Remark: Similar to Example 3, the given conditions are rephrased so that $a$ and $c$ can be expressed in terms of $b$ and then substituted into the algebraic formula. Usually, the result is a polynomial in $b$. But in this example, both the quadratic term and the first-order term of $b$ become 0, so the value of the original algebraic formula is a constant.

If you are familiar with algebraic transformations, take

$$2(a^2 + b^2 + c^2 - ab - bc - ca)$$
$$= (a^2 - 2ab + b^2) + (a^2 - 2ca + c^2) + (b^2 - 2bc + c^2)$$
$$= (a - b)^2 + (a - c)^2 + (b - c)^2.$$

Then, substitute $a - b = 2$, $b - c = 1$, and $a - c = (a - b) + (b - c) = 2 + 1 = 3$ into the computation, and the value of the original formula is $(2^2 + 3^2 + 1^2) \div 2 = 7$.

This method is better than the previous one due to the "equal status" of $a$, $b$, and $c$.

**Example 7.** Given that $a$, $b$, and $c$ are rational numbers, such that $a = 8 - b$ and $c^2 = ab - 16$, find the values of $a$, $b$, and $c$.

**Analysis** This problem requires us to find the values of the letters. We can still use the substitution method: substitute the first formula into the second formula, eliminate $a$, and then examine the result.

**Solution** Substitute $a = 8 - b$ into $c^2 = ab - 16$ to find
$$c^2 = (8 - b)b - 16,$$
that is,
$$c^2 + (b - 4)^2 = 0. \tag{16.1}$$

By $c^2 \geqslant 0$ and $(b - 4)^2 \geqslant 0$, we have $c = 0$ and $b = 4$.
So, $a = 8 - b = 4$.

Remark: $b^2 - 8b + 16 = (b-4)^2$ is simply a square formula. Sometimes, we can add, subtract, or separate some terms to convert a quadratic formula into a square formula plus a constant. For example,

$$x^2 + x + 5 = x^2 + 2 \cdot x \cdot \left(\frac{1}{2}\right) + \left(\frac{1}{2}\right)^2 - \left(\frac{1}{2}\right)^2 + 5$$

$$= \left(x + \frac{1}{2}\right)^2 + \frac{19}{4}.$$

This method is called completing the square.
Here are some more examples.

**Example 8.** Given that $x + \frac{1}{x} = 3$, find the values of

(1) $x^2 + \frac{1}{x^2}$;

(2) $x^3 + \frac{1}{x^3}$.

**Solution**

(1) $x^2 + \frac{1}{x^2} = (x + \frac{1}{x})^2 - 2 = 3^2 - 2 = 7$.
(2)

$$x^3 + \frac{1}{x^3} = \left(x + \frac{1}{x}\right)^3 - 3x \cdot \frac{1}{x}\left(x + \frac{1}{x}\right)$$

$$= 3^3 - 3 \times 3$$

$$= 18.$$

Remark: In this problem, we use the following formulas:

$$(a \pm b)^2 = a^2 + b^2 \pm 2ab,$$
$$(a \pm b)^3 = a^3 \pm b^3 \pm 3ab(a \pm b).$$

**Example 9.** Given that $x + y = 3$ and $xy = 2$, find the values of

(1) $x^3 + y^3$;
(2) $x^4 + y^4$.

**Solution**

(1)

$$
\begin{aligned}
x^3 + y^3 &= (x+y)(x^2 - xy + y^2) \\
&= 3(x^2 + 2xy + y^2 - 3xy) \\
&= 3[(x+y)^2 - 3xy] \\
&= 3 \times (3^2 - 3 \times 2) \\
&= 9.
\end{aligned}
$$

(2)

$$
\begin{aligned}
x^4 + y^4 &= (x^2 + y^2)^2 - 2x^2y^2 \\
&= [(x+y)^2 - 2xy]^2 - 2x^2y^2 \\
&= (3^2 - 2 \times 2)^2 - 2 \times 2^2 \\
&= 17.
\end{aligned}
$$

Remark: The symmetric polynomials of $x$ and $y$ (that is, polynomials that remain unchanged after $x$ and $y$ are interchanged, such as $x^3 + y^3$ and $x^4 + y^4$) can be expressed as $x + y$ and $xy$.

This problem is actually the case of $x + y = 3$ and $xy = 2$.

In this problem, $x = 1$, $y = 2$ or $x = 2$, $y = 1$ can be obtained from the given equations, and then the values of $x^3 + y^3$ and $x^4 + y^4$ can be obtained. But in most cases, the values of $x$ and $y$ are not integers, so the computation will become complicated.

**Example 10.** Given that $\frac{a}{a^2+a+1} = \frac{1}{6}$. Try to find the value of $\frac{a^2}{a^4+a^2+1}$.

**Analysis** The numerator is simpler than the denominator, so we take the reciprocal of the fraction.

**Solution**   From the given condition, $a \neq 0$ and $\frac{a^2+a+1}{a} = 6$, that is,

$$a + \frac{1}{a} = 5.$$

So,

$$\frac{a^4 + a^2 + 1}{a^2} = a^2 + \frac{1}{a^2} + 1 = \left(a + \frac{1}{a}\right)^2 - 1$$

$$= 5^2 - 1 = 24.$$

Therefore, the original formula $= \frac{1}{24}$.


**Exercises**

1. According to the following values of $a$ and $b$, find the values of the algebraic formula $a^2 - \frac{b^2}{a}$:

   (1) $a = 2$ and $b = 1$;
   
   (2) $a = \frac{2}{3}$ and $b = \frac{1}{9}$.

2. Given that $x = (-1 \div \frac{1}{2} \times 3 \times \frac{1}{6})^3$, find the value of the algebraic formula $x^{2007} + x^{2006} + x^{2005} + \cdots + x + 1$.

3. The value of $ax^3 + bx + 8$ is 12 when $x = 3$. Find the value of $ax^2 + bx - 5$ when $x = 3$.

4. If $\frac{x}{3} = \frac{y}{4} = \frac{z}{5}$ and $4x - 5y + 2z = 10$, find the value of $2x - 5y + z$.

5. Given that $a + b = 3$ and $ab = 2$, find the value of $a^2 + b^2$.

6. If $x + y = 1$, then find the value of $x^3 + y^3 + 3xy$.

7. If $3x^2 - x = 1$, then find the value of the algebraic formula $6x^3 + 7x^2 - 5x + 2006$.

8. Given that $x^2 - 3x + 1 = 0$, find the following values:

   (1) $x^2 + \frac{1}{x^2}$;
   
   (2) $x^3 + \frac{1}{x^3}$;
   
   (3) $x^4 + \frac{1}{x^4}$.

9. Let $y = ax^4 + bx^2 + c$. When $x = -5$, we have $y = 3$. Find the value of $y$ when $x = 5$.

10. Suppose that $(2x - 1)^5 = a_5x^5 + a_4x^4 + a_3x^3 + a_2x^2 + a_1x + a_0$. Find the following values:

    (1) $a_0 + a_1 + a_2 + a_3 + a_4 + a_5$;
    
    (2) $a_0 - a_1 + a_2 - a_3 + a_4 - a_5$;
    
    (3) $a_0 + a_2 + a_4$.

# Chapter 17

# Logical Inference I

Some problems do not require a lot of mathematical knowledge or calculations, and conclusions are drawn mainly using common sense and logic. Such problems are called logical inference problems.

**Example 1.** A lion lies on Mondays, Tuesdays, and Wednesdays. A unicorn lies on Thursdays, Fridays, and Saturdays. They tell the truth on other days, respectively. The son of the forest asked the lion, "What day is it today?" The lion said, "I lied yesterday." He asked the unicorn the same question, and the unicorn said, "Yesterday was the day I lied." Do you know what day it is today?

**Solution** The lion lies on Mondays, so he can say on Mondays, "I lied yesterday." Actually, Sunday is not a day he lies.

And on Tuesdays and Wednesdays, the lion doesn't tell the truth, so he doesn't say, "I lied yesterday."

On Thursday, the lion tells the truth, so he can say, "I lied yesterday."

On Fridays, Saturdays, and Sundays, the lion speaks the truth and will not say, "I lied yesterday."

Therefore, the lion said, "I lied yesterday," indicating that today is Monday or Thursday.

In the same way, the unicorn said, "Yesterday was the day I lied," indicating that today is Thursday or Sunday.

Only Thursday matches both conditions. So, according to what they said, today is Thursday.

Remark: The methods used for solving this problem are enumeration and reduction to absurdity.

**Example 2.** $A$, $B$, $C$, $D$, $E$, and $F$ are six suspects. When they were interrogated, their confessions were as follows:

$A$: "$B$ and $F$ committed the crime."
$B$: "$D$ and $A$ committed the crime."
$C$: "$B$ and $E$ committed the crime."
$D$: "$A$ and $C$ committed the crime."
$E$: "$F$ and $A$ committed the crime."

Given that the crime was committed by two people in partnership, one of the above confessions is a lie, and the other four are half-true and half-false. Which two people are the criminals?

**Solution** According to the conditions, one of the five confessions is false, and the remaining four are half-true and half-false. Exactly one of the names is a criminal in these four confessions. The names of the two criminals appear totally 4 times in the five confessions.

In all confessions, $A$ appears 3 times, $B$ appears 2 times, $C$ appears 1 time, $D$ appears 1 time, $E$ appears 1 time, and $F$ appears 2 times. Therefore, it can only be one of $A$ and $C$, $A$ and $D$, $A$ and $E$, or $B$ and $F$ to commit the crime.

If $A$ and $C$ commit the crime together, then $D$'s confession is all true, which contradicts the given condition.

If $A$ and $D$ commit the crime together, then $B$'s confession is all true, which contradicts the given condition.

If $B$ and $F$ commit the crime together, then $A$'s confession is all true, which contradicts the given condition.

Therefore, it can only be that $A$ and $E$ committed the crime together; that is, the criminals are $A$ and $E$.

Remark:   To solve this kind of logical inference problem, it is very important to find a breakthrough point. For example, in this problem, the names of the criminals appear four times in total, which is a breakthrough point.

**Example 3.** Three basketball teams, $A$, $B$, and $C$, have one basketball game every day. It is stipulated that the winning team will play against the other team on the next day after each game, while the loser will have a day off. If the final result is that team $A$ won 10 games, team $B$ won 12 games, and team $C$ won 14 games, how many games did each team play?

**Solution** First, we compute the total number of games they played, and then we study the relationship between the number of games lost by a certain team and the number of rest days.

According to the rules of the basketball games, we know that each game will definitely produce a winner and a loser, and hence the sum of all teams' wins is the total number of games: there is a total of $10 + 12 + 14 = 36$ games, which is 36 days. Suppose the teams $A$, $B$, and $C$ lose $a_1$, $b_1$, and $c_1$ games and rest on $a_2$, $b_2$, and $c_2$ days, respectively, as shown in the following table:

| Teams | Win | Lose | Rest | Total |
|-------|-----|------|------|-------|
| $A$ | 10 | $a_1$ | $a_2$ | 36 |
| $B$ | 12 | $b_1$ | $b_2$ | 36 |
| $C$ | 14 | $c_1$ | $c_2$ | 36 |

Now, consider team $A$. The key observation is that each loss is followed by a rest, except for a rest on the first day. There are two situations:

(1) $A$ has a rest on the first day. If $A$ does not lose on the last day, then $a_2 = a_1 + 1$; if $A$ loses on the last day, then $a_2 = a_1$.
(2) $A$ does not have a rest on the first day. If $A$ does not lose on the last day, then $a_2 = a_1$; if $A$ loses on the last day, then $a_2 = a_1 - 1$.

So, $a_2 = a_1 + 1$, $a_2 = a_1 - 1$, or $a_2 = a_1$.

As $10 + a_1 + a_2 = 36$, so $a_1 + a_2 = 26$ is an even number, implying that $a_1$ and $a_2$ have the same parity and they cannot differ by 1, we have $a_1 = a_2 = 13$.

For the same reason, we have $b_1 = b_2 = 12$ and $c_1 = c_2 = 11$.

So, team $A$ played $10 + a_1 = 23$(games), team $B$ played $12 + b_1 = 24$(games), and team $C$ played $14 + c_1 = 25$ (games).

Remark: In this problem, parity plays an important role.

The following Example 4 is a bit long and needs to be read carefully.

**Example 4.** Four people, $A$, $B$, $C$, and $D$, live in an 18-floor building. Among them are a professor, an engineer, a doctor, and a student. Given that:

(a) $D$ lives above $A$, and $A$ lives above $C$.
(b) $B$ lives under the doctor, and the doctor lives under the professor.
(c) if the number of the floor the engineer lives on increases by 2, then the number of floors between him and the doctor is exactly the same as the number of floors between him and the professor.
(d) if the number of the floor where the engineer lives on is reduced by half, then he lives exactly in the middle of the floors where the student

and the doctor live (that is, the average of the floor numbers where the student and the doctor live).

Find the jobs of $A$, $B$, $C$, and $D$.

If we know further that:

(e) the number of the floor $D$ lives on is exactly five times the number of the floor the student lives on, try to find the number of floors each person lives on.

## Solution

(a) indicates that the order from top to bottom is $D$, $A$, and $C$.

(b) indicates that the order from top to bottom is the professor, the doctor, and $B$.

(d) does not specify the relative positions of the doctor and the student, but the number of the floor where the engineer lives on divided by 2 is exactly between the floors where the student and the doctor live; that is, the number of the floor he lives on is even and equal to the floor number of the student plus the floor number of the doctor. This shows that the engineer lives above the doctor and the student.

Both the professor and the engineer live above the doctor, and the doctor lives above $B$, so $B$ is the student.

Therefore, $C$ is the doctor.

From (c), the engineer lives under the professor, so $A$ is the engineer and $D$ is the professor.

Suppose that $A$, $B$, $C$, and $D$ live on floors $a$, $b$, $c$, and $d$, respectively. Then, as mentioned above, $a$ is an even number and

$$a = b + c. \tag{17.1}$$

By (c), we have

$$2(a + 2) = d + c. \tag{17.2}$$

By (e), we have $d = 5b$, so it is easy to get $d = 15$ and $b = 3$.

Therefore, eliminating $c$ from (17.1) and (17.2), we have

$$a = d - b - 4 = 15 - 3 - 4 = 8;$$

$$c = a - b = 8 - 3 = 5.$$

**Answer**   $D$ is the professor, $A$ is the engineer, $C$ is the doctor, and $B$ is the student, and they live on floors 15, 8, 5, and 3, respectively.

Remark:   Do not miss any information provided by the problem, such as $a$ being an even number. Paying attention to this can save a lot of unnecessary discussions.

**Example 5.** In a certain exam, there are six questions in total, all of which are true or false questions. If the candidate thinks it is true, he will draw "O," and if he thinks it is false, he will draw "X." The method of scoring is: 2 points for each question answered correctly, 1 point for each question unanswered, and no points for wrong answers.

Given the answers of Zhao, Qian, Sun, Li, Zhou, Wu, and Zheng and the scores of the first six people in the following table, fill in Zheng's score in the table.

| Question \ Name | Zhao | Qian | Sun | Li | Zhou | Wu | Zheng |
|---|---|---|---|---|---|---|---|
| 1 | | O | O | O | X | X | O |
| 2 | O | | X | X | O | X | X |
| 3 | O | X | | O | X | X | X |
| 4 | O | O | X | | X | O | O |
| 5 | O | X | O | O | | X | O |
| 6 | O | O | X | X | X | | X |
| Score | 7 | 5 | 5 | 5 | 9 | 7 | |

**Solution**   It can be seen from the table that Zhou has the highest score of 9 points, and only one of the five questions he answered was wrong. If we know which question he got wrong, then the correct answer to each question can be deduced, so we choose him as the breakthrough.

Direct enumeration is cumbersome; we can make some comparisons with scores in the table first.

Sun and Li have the same total score but different scores for questions 3 and 4 (scores for the rest of the questions are the same). Because they got 1 point for each unanswered question, the score for question 4 of Sun is the same as the score for question 3 of Li, which is either 2 or 0 (otherwise the two people will not have the same score). Either way, there is exactly one correct answer to questions 3 and 4 of Zhou. Therefore, questions 1, 2, and 6 were answered correctly by Zhou.

Comparing Zhao's and Zhou's answers, Zhao's first four questions scored $1 + 2 + 2 = 5$ (exactly one of questions 3 and 4 is correct), and question 6 got 0 points, so Zhao got $7 - 5 = 2$ (points) for question 5; that is, the answer to question 5 is true.

Comparing the answers of Wu, Zhou, and Zhao, Wu got $2+0+0+1 = 3$ (points) for questions 1, 2, 5, and 6, so he got $7 - 3 = 4$ (points) for questions 3 and 4. That is, his answers to both questions 3 and 4 were correct.

So, the answers to these six questions are X, O, X, O, O, and X. Zheng answered the last four questions correctly and got $2 \times 4 = 8$ (points).

Remark: Of course, the method is not unique, as there are many ways to compare the answers and scores. The following method is also recommended: for each question, among the first 6 people, there are always 3 whose answers are the same as Zheng's, 2 whose answers are different from Zheng's, and 1 blank. If Zheng answered correctly, then the first 6 people would get 7 points; if Zheng was wrong, then the first 6 people would get 5 points. In any case, for each question, they would get 5 points more than Zheng. Therefore, the total score of the first 6 people is $5 \times 6 = 30$ (points) more than Zheng. So, Zheng's score is $7+5+5+5+9+7-30 = 8$ (points).

This solution does not involve the correct answer to each question but directly gives Zheng's score. It has advantages and disadvantages (if we want to know the answer to each question).

For logical inference problems, not much mathematical knowledge is used, and we mainly rely on the analysis of the given conditions to find appropriate breakthroughs. Methods such as enumeration and reduction to absurdity are commonly used. But for each specific problem, we need to come up with our own ideas to find a suitable solution.

## Reading

Have you read Shakespeare's famous play, "The Merchant of Venice"?

The mathematician R. Smullyan added a plot to the play:

Antonio went to Portia's house to propose to her. Portia took out three boxes, one gold, one silver, and one aluminum and said, "There is a sentence written on each box, but only one sentence is true. Whoever can guess which box my portrait is in will be my husband."

The boxes can be checked one by one using the enumeration method.

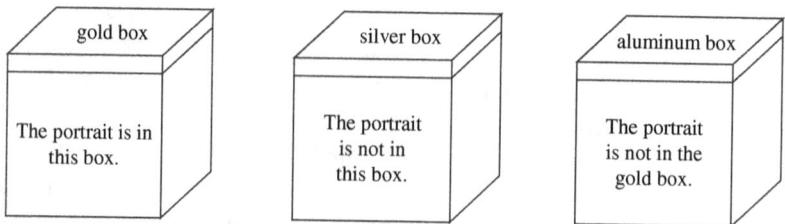

Fig. 17.1

If the portrait is in the gold box, then the words on the gold and silver boxes are both true, contradicting the fact that there is only one true statement.

If the portrait is in the silver box, only the words on the aluminum box are true, which is consistent with Portia's announcement.

If the portrait is in the aluminum box, then the words on the silver and aluminum boxes are both true, which contradicts the fact that there is only one true statement.

So, the portrait is in the silver box.

The wise and honest Antonio certainly guessed the correct answer.

There is another solution to this question. Note that the words on the gold box are exactly opposite those on the aluminum box. Therefore, exactly one of the two sentences is true. As there is only one true statement, the words on the silver box are false; that is, the portrait must be in the silver box.

---

## Exercises

1. In a class, 30 students participated in the sports team, 25 students participated in the art team, and 13 students participated in both teams. Each student participated in at least one team. Find the total number of students in the class.

2. Someone said, "I have been lying since last year and never told the truth." Is this sentence true?

3. A visiting group chooses some places to visit from five cities $A$, $B$, $C$, $D$, and $E$ according to the following restrictive conditions:

   (1) if they go to city $A$, they must also go to city $B$;
   (2) they will go to at least one of cities $D$ and $E$;
   (3) they will go to only one of cities $B$ and $C$;
   (4) they either go to both cities $C$ and $D$, or go to neither of cities $C$ and $D$;
   (5) if they go to city $E$, then they must go to cities $A$ and $D$.

   Try to analyze at most how many cities the visiting group can go to and what are they?

4. Each of three teachers, Li, Wang, and Zhang, is responsible for two of the six subjects of biology, physics, English, physical education, history, and mathematics. It is known that: ① the physics teacher and the physical education teacher are neighbors; ② teacher Li is the youngest

among the three; ③ teacher Zhang, the biology teacher, and the physical education teacher often go home from school together; ④ the biology teacher is older than the math teacher; ⑤ the English teacher, mathematics teacher, and Teacher Li all love playing volleyball.

Based on the above information, determine which two courses they are responsible for.

5. Students $A$, $B$, $C$, and $D$ have all solved a problem correctly. The teacher asked them, "who solved it first?" They all responded humbly as follows:

$A$: "It's not me."
$B$: "It's $D$."
$C$: "It's $B$."
$D$: "It's not me."

Of the four answers, only one is true. Who actually solved the problem first?

6. People $A$, $B$, $C$, $D$, and $E$ each borrow a novel from a library and agree to exchange after reading their respective books. The thickness of the five books and the reading speed of these five people are the same, so they always exchange at the same time. After four exchanges, everyone has read all five books. Suppose the following conditions:

(1) The last book $A$ read is the second book $B$ read.
(2) The last book $C$ read is the fourth book $B$ read.
(3) The second book $C$ read is the first book $A$ read.
(4) The last book $D$ read is the third book $C$ read.
(5) The fourth book $B$ read is the third book $E$ read.
(6) The third book $D$ read is the first book $C$ read.

Who first read the second book that $D$ read?

7. There are 27 pearls, one of which is fake, but its appearance is the same as a real one and only a little lighter than a real one. All the real pearls have the same weight. Can you find out the fake pearl by weighing three times with a balance (without weights)?

8. One of Wang, Chen, and Zhang did something bad, and the other two know who did it. When the teacher asked about the situation, the three of them said the following sentences:

Chen: "I didn't do it. Zhang didn't do it either."
Wang: "I didn't do it. Chen didn't do it either."
Zhang: "I didn't do it. I don't know who did it."

Given that each of them told a lie and a truth, who did the bad thing?

9. (1) In the group stage of football world cup, four teams in each group will play a single round-robin match. In each game, the winning team will get 3 points, the losing team will get 0 points, and the two teams will get 1 point for a tie. After the group stage is over, the two teams with the highest total points qualify for the next round. If the total points are the same, they will be sorted according to the goal differences and the numbers of goals scored. How many points will ensure a team qualifies for the next round? Briefly explain the reason.

   (2) In the group stage, is it possible for a team to qualify with only 3 points? What about only 2 points?

10. Among three students, Hong, Qiang, and Hua, one of them volunteered to clean the classroom. Afterward, the teacher asked them, "who did the good thing?"

   Hong said: "Qiang did it."
   Qiang said: "I didn't do it."
   Hua said: "I didn't do it."

   If two of them told lies and one told the truth, can the teacher determine who did it?

# Chapter 18

# Logical Inference II

We introduce some more logical inference problems in this chapter.

**Example 1.** A total of 160 people stand in a line and report in order from 1 to 160. Then, all the odd-numbered people leave. Those who stay behind count from 1, and the odd-numbered ones leave again. This will continue until, finally, one person is left. What is the number of this person's first report?

**Solution** Every positive integer can be written in the form of $2^s(2k+1)$, where $s$ and $k$ are 0 or positive integers.

After the first report, those people who reported odd numbers left; that is, all the people whose reported number with $s = 0$ left.

In the second report, the person whose first reported number was $2^s(2k+1)$ now reports $2^{s-1}(2k+1)$. After reporting, all the people whose first reported number with $s = 1$ left.

By analogy, the last person left had the largest $s$ value among all the numbers first reported. As

$$2^7 < 160 < 2^8,$$

for the last person left, his first reported number is $2^7$, which is 128.

**Example 2.** There are two decks of playing cards. The order of arrangement of each deck is as follows: the first card is the red joker, the second is the black joker, and then there are four suits of spades, hearts, diamonds, and clubs. The cards of each suit are arranged in the order of A, 2, 3, ..., J, Q, and K. Someone puts the two decks of playing cards arranged together one above the other, and then removes the first card and puts the

161

second card at the bottom, then removes the third card, and then puts the fourth card at the bottom, and so on, until there is only one card left. Which card is the remaining card?

**Solution** Label the cards with numbers 1–108. In the first round, 54 cards are removed, and the remaining 54 cards are all even numbers. In the second round, 27 cards are removed, and the remaining numbers are all multiples of 4.

Note that 27 is an odd number. In the third round, 14 cards are removed, and 13 cards are left. The last card taken away in the third round is $4 \times 27 = 108$. The rest are 8, 16, 24, 32, 40, 48, 56, 64, 72, 80, 88, 96, and 104. Next, 16 (not 8 because 8 comes after 108), 32, 48, 64, 80, 96, 8, 40, 72, 104, 56, and 24 will be taken in turn. The last one left is 88.

$$88 - 54 - 2 - 13 - 13 = 6.$$

That is, the last remaining card is the 6 of diamond.

Remark: Note the similarities and differences between Example 2 and Example 1.

**Example 3.** A school held a mathematics competition, and five students, $A$, $B$, $C$, $D$, and $E$, achieved the top five positions. Before awarding the prizes, the teacher asked them to guess their rankings.

$A$ said: "$B$ is third and $C$ is fifth."
$B$ said: "$E$ is fourth and $D$ is fifth."
$C$ said: "$A$ is first and $E$ is fourth."
$D$ said: "$C$ is first and $B$ is second."
$E$ said: "$A$ is third and $D$ is fourth."

The teacher said: "Every ranking has been guessed correctly by someone among you." What are the rankings of these five students?

**Solution** According to the problem conditions, we can list a table as follows:

| Guesser \ Ranking | First | Second | Third | Fourth | Fifth |
|---|---|---|---|---|---|
| $A$ |  |  | $B$ |  | $C$ |
| $B$ |  |  |  | $E$ | $D$ |
| $C$ | $A$ |  |  | $E$ |  |
| $D$ | $C$ | $B$ |  |  |  |
| $E$ |  |  | $A$ | $D$ |  |

It can be seen from the above table that $B$ is the only person who is guessed as ranked in the second place, and there are students who guessed correctly for every position, so it can be concluded that $B$ is ranked in the second place, not the third place. From this, it can be deduced that $A$ is ranked in the third place. Since $A$ is not in the first place, we have that $C$ is ranked in the first place and then $D$ is in the fifth place. Finally, $E$ is ranked in the fourth place.

Therefore, the rankings of these five students are as follows: $C$ is the first, $B$ is the second, $A$ is the third, $E$ is the fourth, and $D$ is the fifth.

**Remark:** By listing a table, we can make the conditions and internal relations of some logical inference problems clearer. Some parts of the problem are like dominoes. As long as the breakthrough is found, the problem will be quickly solved by the domino effect.

**Example 4.** $A$, $B$, $C$, and $D$ play a chess tournament, and each two of them compete in a game. As a result, $B$ beats $D$; $A$, $B$, and $C$ win the same number of games. If there are no ties in the tournament, how many games does $D$ win?

**Solution** There are totally 6 games: $AB$, $CA$, $AD$, $BC$, $BD$, and $CD$. According to the given conditions, $D$ has lost to $B$, and so the number of games that $D$ wins is less than 3.

If $D$ wins 1 game, then $A$, $B$, and $C$ win 5 games in total.

If $D$ wins 2 games, then $A$, $B$, and $C$ win 4 games in total.

As neither 4 nor 5 is a multiple of 3, neither of these two situations can make $A$, $B$, and $C$ win the same number of games. Therefore, $D$ only wins 0 games, and $A$, $B$, and $C$ each win 2 games.

**Example 5.** People's blood types are usually A, B, O, and AB. The relationship between the blood types of children and their parents is shown in the following table:

| Blood type of parents | Possible blood type of children |
| --- | --- |
| O, O | O |
| O, A | A, O |
| O, B | B, O |
| O, AB | A, B |
| A, A | A, O |
| A, B | A, B, AB, O |
| A, AB | A, B, AB |
| B, B | B, O |
| B, AB | A, B, AB |
| AB, AB | A, B, AB |

There are three children wearing red, yellow, and blue jackets, whose blood types are O, A, and B in order. For each child, his/her parents' hats have the same color. The colors of the hats are also red, yellow, and blue, representing the blood types of AB, A, and O in order. For each child, find the color of his/her parents' hats.

**Analysis** As the parents of each child wear hats of the same color, the blood types of the parents are the same. In this way, only rows 1, 5, 8, and 10 of the table are relevant to the problem. As no hat color represents type B, row 8 can also be crossed out. In this way, the blood type table becomes much smaller than before, and the discussion is much simpler, making it easy to get the answers to this problem.

**Solution** As both parents of each child wear hats of the same color, the blood types of both parents are the same, and there is no type B blood. We only need to consider the following short table:

| Blood type of parents | Possible blood type of children |
|:---:|:---:|
| O, O | O |
| A, A | A, O |
| AB, AB | A, B, AB |

From the above short table, we can find that if the parents' blood type is O, the child's blood type must be O; that is, if the child wears a red jacket, the parents must wear blue hats.

Cross out the blood type O of all the children and the first row of the short table.

From the second line, if the parent's blood type is A, then the child's blood type must be A; that is, if the child wears a yellow jacket, then the parents wear yellow hats.

Finally, for the child wearing a blue jacket, his parents wear red hats.

**Example 6.** Fill distinct (no two are equal) positive integers into an $18 \times 18$ table, one number in every square. Prove that no matter how we fill the numbers, there are at least two pairs of adjacent squares (two squares with a common side are called adjacent) such that for each pair, the difference between the numbers filled in the adjacent squares is at least 10.

**Proof** Suppose that $a$ and $b$ are the smallest and largest numbers of these 324 positive integers, respectively.

As these numbers are distinct from each other, $b - a \geqslant 323$.

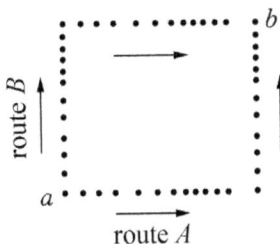

Fig. 18.1

(1) When the squares with $a$ and $b$ are in different rows and different columns, starting from the square with $a$, we can reach the square with $b$ by moving horizontally and vertically, as shown in Figure 18.1. As $a$ and $b$ are in different rows and different columns, the two routes $A$ and $B$ have only two squares in common, namely the squares with $a$ and $b$. Obviously, each route passes through at most $18 + 17 = 35$ squares, or 34 pairs of adjacent squares.

If the difference between the numbers filled in every two adjacent squares on route $A$ is less than or equal to 9, then we have

$$323 \leqslant b - a \leqslant 34 \times 9 = 306.$$

A contradiction! Therefore, there must be two adjacent squares on route $A$ such that the difference between the numbers filled in them is greater than or equal to 10; the same is true for route $B$.

(2) When the squares with $a$ and $b$, are in the same row or in the same column.

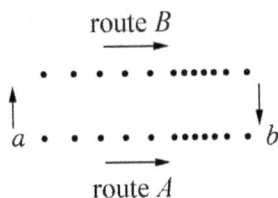

Fig. 18.2

Similar to case (1), we can also find two completely different routes $A$ and $B$, each passing through 35 or fewer squares, as shown in Figure 18.2: route $A$ consists of squares that are in the same row (or in the same column) as $a$ and $b$; route $B$ consists of $a$, $b$, and the row above them (or to the left or right column if they are in the same column). There is a pair of adjacent squares in each of the two routes, and the difference between the numbers filled in them is greater than or equal to 10.

## Exercises

1. Four children, $A$, $B$, $C$, and $D$, were playing football in the yard, when they broke the window glass of a nearby room. When asked, they gave the following answers:

   $A$ said: "$B$ did it." $B$ said: "$D$ did it." $C$ said: "I didn't break it." $D$ said: "$B$ lied."

   Given that only one of the children told the truth and only one of them broke the window glass, then who did tell the truth and who broke the window glass?

2. There is a cube, each face of which is marked with a letter, $a, b, c, d, e,$ or $f$. Students $A$, $B$, and $C$ observe the cube from three different angles, and the observation results are shown in Figure 18.3. What are the three letters on the opposite faces of the cube?

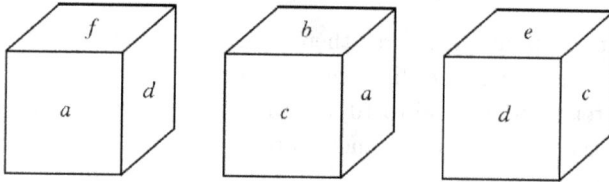

Fig. 18.3

3. Beads of five colors, red, blue, yellow, white, and purple, are wrapped in paper bags and arranged in a row on a table. Five people guess the color of the bead in each paper bag.

   $A$'s guess: "the bead in the second bag is purple and the bead in the third bag is yellow."

   $B$'s guess: "the bead in the second bag is blue and the bead in the fourth bag is red."

   $C$'s guess: "the bead in the first bag is red and the bead in the fifth bag is white."

   $D$'s guess: "the bead in the third bag is blue and the bead in the fourth bag is white."

*E*'s guess: "the bead in the second bag is yellow and the bead in the fifth bag is purple."

After guessing, they open the paper bags and take a look. They find that everyone has guessed one color correctly, and each bag has been guessed by one person correctly. Which color bead did each person guess correctly?

4. *S*, *J*, and *R* are the brakeman, stoker, and driver (not necessarily in that order) on a train. There are only three passengers on the train today, and it just so happens that the last names of the three passengers are also *S*, *J*, and *R*. To distinguish the staff from the passengers, let's refer to the passengers as Mr. *S*, Mr. *J*, and Mr. *R*. Additionally, we also know that:

   (1) Mr. *R* lives in Detroit;
   (2) the brakeman lives somewhere between Chicago and Detroit;
   (3) the passenger living in Chicago has the same last name as the brakeman;
   (4) a neighbor of the brakeman is also a passenger and his annual salary is exactly three times that of the brakeman (each annual salary is an integer);
   (5) Mr. *J* earns exactly 20000 yuan a year and has to live on government relief;
   (6) Mr. *S* plays billiards better than the stoker;

   Now, the question is, who is the driver?

5. The school offers four extracurricular courses in Chinese, Mathematics, English, and Science for students to enroll in voluntarily. The numbers of students participating in the Chinese, Mathematics, English, and Science interest courses in a class are 18, 20, 21, and 19, respectively. If the total number of students in the class is 25, at least how many students in the class have signed up for all four interest courses?

6. Each of *A*, *B*, and *C* has a number of candies in their hands. In the first round, *A* matches the numbers of candies in *B*'s and *C*'s hands and gives the candies to them, respectively. Similarly, in the second and third rounds, *B* and *C* match the candies in the other two person's hands and give the candies to them, respectively. (So, after each round, two persons' candies are doubled.) At this time, each of *A*, *B*, and *C* has 8 candies in their hands. How many candies did *A*, *B*, and *C* have originally?

7. When discussing an open test question, the teacher required each student to exchange opinions with at least three students in the same group. There were 11 students in a study group. After the discussion, two students said that they exchanged opinions with 4 students. Please prove that at least one other student also exchanged opinions with at least 4 students.

8. There are a total of $n(n \geqslant 3)$ players participating in a sports competition, and every two players compete for one game. Suppose that there are no draws and each player fails to beat all his/her opponents. Prove that among them, there must be 3 players, $A$, $B$, and $C$, such that $A$ beats $B$, $B$ beats $C$, and $C$ beats $A$.

9. Zhang, Li, and Wang were born in Beijing, Shanghai, and Wuhan. Their occupations are singer, cross-talk actor, and dancer. Suppose that:

   (1) Wang is not a singer and Li is not a cross talk actor;
   (2) the singer was not born in Shanghai;
   (3) the cross-talk actor was born in Beijing;
   (4) Li was not born in Wuhan.

   Try to determine the occupations and birthplaces of Wang, Li, and Zhang.

10. Write three identical positive integers on a blackboard, and then erase one of them and replace it with the difference between the sum of the other two numbers and 1. Repeat this process several times to get $(17, 1983, 1999)$. What are the three numbers written on the blackboard at the beginning?

# Chapter 19

# Divisibility

For positive integers $a$ and $b$ (all the letters in this chapter represent integers), we have division with remainder:

$$a = bq + r (0 \leqslant r < b),$$

where $q$ is called the quotient and $r$ is called the remainder.

In particular, if $r = 0$ and $a = bq$, then $a$ is said to be divisible by $b$, or $b$ divides $a$, which is denoted as $b \mid a$. Now, we also say that $a$ is a multiple of $b$ and $b$ is a divisor (or factor) of $a$.

If $r \neq 0$, then we say that $b$ does not divide $a$, denoted as $b \nmid a$.

As $0 = 0 \times b$, so we treat 0 as a multiple of any integer $b$.

For general integers $a$ and $b$ ($b \neq 0$), if $|b| \mid |a|$, then $b \mid a$.

Usually, when we talk about multiples and divisors, we mean positive multiples and positive divisors.

Divisibility has the following basic properties:

1. If $a \mid b$ and $b \mid c$, then $a \mid c$.
2. If $a \mid b$ and $k$ is an integer, then $a \mid kb$.
3. If $a \mid b$ and $a \mid c$, then $a \mid (b \pm c)$.
4. If $a \mid bc$ and $a$ is co-prime with $c$, then $a \mid b$.
   Especially, if a prime number $p \mid bc$, then $p \mid b$ or $p \mid c$.
5. If $b \mid a$, $c \mid a$, and $b$ is co-prime with $c$, then $bc \mid a$.

The above properties, especially 4 and 5, can be gradually understood through the following examples.

**Example 1.** How many positive integers within 100 are divisible by 2, 3, and 5 at the same time?

**Solution**  As 2 and 3 are co-prime, 3 and 5 are co-prime and 5 and 2 are co-prime (this characteristic can also be stated as that 2, 3, and 5 are relatively prime), according to property 5, an integer that is divisible by 2, 3, and 5 at the same time must be divisible by $2 \times 3 \times 5 = 30$. On the other hand, by property 1, positive integers that are divisible by 30 must be divisible by 2, 3, and 5 at the same time. Therefore, within 100, those divisible by 2, 3, and 5 at the same time are just multiples of 30. Obviously, there are only three such numbers: 30, 60, and 90.

Remark:   If $a_1, a_2, \ldots, a_m$ are pairwise co-prime, then the positive integers that are divisible by $a_1, a_2, \ldots, a_m$ are the positive integers that are divisible by $a_1 \cdot a_2 \cdots \cdots a_m$, which is a very useful conclusion.

Generally, when $a_1, a_2, \ldots, a_m$ are not pairwise co-prime, the following conclusions can be drawn:

A positive integer that is divisible by $a_1, a_2, \ldots, a_m$ at the same time is an integer that is divisible by the least common multiple of $a_1, a_2, \ldots, a_m$.

In addition, to find the number of positive integers that are divisible by 30 within 100, it is not necessary to list all 30, 60, and 90; rather, we divide 100 by 30, then the quotient 3 is the number we want to find.

**Example 2.** Find the number of positive integers within 1000 that are divisible by 3, 4, 5, and 6 at the same time.

**Solution**  The least common multiple of 3, 4, 5, and 6 is 60. So, within 1000, the positive integers that are divisible by 3, 4, 5, and 6 at the same time are just the positive integers that are divisible by 60. As the quotient of 1000 divided by 60 is 16, there are 16 such positive integers.

**Example 3.** Prove that six-digit numbers of the form $\overline{abcabc}$ must be divisible by 7, 11, and 13.

**Proof**  $\overline{abcabc} = \overline{abc} \times 1001 = \overline{abc} \times 7 \times 11 \times 13$.

Then, we have that $\overline{abcabc}$ is divisible by 7, 11, and 13.

**Example 4.** Suppose that the five-digit number $\overline{x679y}$ is divisible by 72. Find the digits $x$ and $y$.

**Solution**  As $72 = 8 \times 9$, $\overline{x679y}$ is divisible by 8 and 9.

When a number is divisible by 8, the number composed of the last three digits of this number must be divisible by 8. So, $\overline{79y}$ is divisible by 8. Through division, it is not difficult to find $y = 2$.

When a number is divisible by 9, its digit sum is divisible by 9, so $x + 6 + 7 + 9 + 2 = x + 24$ is divisible by 9. Note that $0 < x \leqslant 9$; therefore, $x$ can only be 3. So, $x = 3$ and $y = 2$.

Remark: We use some characteristics of integers that are divisible by 8 and 9 in this example to successfully solve the problem. This type of problem often appears in competitions. In the following, we give some commonly used divisibility characteristics:

(1) numbers that are divisible by 2: the ones digit is an even number;
(2) numbers that are divisible by 5: the ones digit is 0 or 5;
(3) numbers that are divisible by 4: the number composed of the last two digits is divisible by 4;
(4) numbers that are divisible by 25: the number composed of the last two digits is divisible by 25;
(5) numbers that are divisible by 8: the number composed of the last three digits is divisible by 8;
(6) numbers that are divisible by 125: the number composed of the last three digits is divisible by 125;
(7) numbers that are divisible by 3: the digit sum of this number is divisible by 3;
(8) numbers that are divisible by 9: the digit sum of this number is divisible by 9;
(9) numbers that are divisible by 11: the difference between the sum of odd-numbered digits and the sum of even-numbered digits is divisible by 11.

More generally, we have the following:

When a number is divided by 2 or 5, the remainder is the same as the remainder of the ones digit divided by 2 or 5.

When a number is divided by 4 or 25, the remainder is the same as the remainder of the number composed of the last two digits of this number divided by 4 or 25.

When a number is divided by 8 or 125, the remainder is the same as the remainder of the number composed of the last three digits of this number divided by 8 or 125.

When a number is divided by 3 or 9, the remainder is the same as the remainder of the digit sum of this number divided by 3 or 9.

When a number is divided by 11, the remainder is the same as the remainder of the difference obtained by subtracting the sum of its odd-numbered digits (starting from the right) and the sum is given by the even-numbered digits divided by 11.

**Example 5.** $N = \underbrace{19991999\cdots1999}_{1999 \text{ of } 1999}$. Find the remainder of $N$ divided by 11.

**Solution**  Obviously, the difference between the sum of its odd-numbered digits and the sum of its even-numbered digits is

$$1999 \times (9 + 9 - 9 - 1) = 1999 \times 8.$$

The remainder of $1999 \times 8$ divided by 11 is the same as the remainder of $8 \times 8$ divided by 11, that is, the remainder is 9. Therefore, the remainder of $N$ divided by 11 is 9.

**Example 6.** Line up 2002 students in a row, and number them from 1 to 2002 from left to right. Then, they report repeatedly from 1 to 11 from left to right. The students who report 11 stay where they are, while the rest leave the line. The remaining students report repeatedly from 1 to 11 from left to right again, and the students who report 11 stay, while the rest leave. The remaining students report from 1 to 11 from left to right for the third time; the students who report 11 stay, while the rest leave. How many students are left at the end? What are their numbers?

**Solution**  According to the problem conditions, for the remaining students after the first report, their numbers must be a multiple of 11.

For the remaining students after the second report, their numbers must be a multiple of $11^2 = 121$.

For the remaining students after the third report, their numbers must be a multiple of $11^3 = 1331$.

Therefore, for the students left at the end, their numbers must be a multiple of 1331. As we know that from 1 to 2002, there is only one multiple of 1331, that is, 1331. So, there is only one student left at the end, and his number is 1331.

**Example 7.** Write 13 consecutive natural numbers that are all composite.

**Solution**  If a natural number $a$ is a multiple of 2, then $a + 2$ is also a multiple of 2. If $a$ is a multiple of 3, then $a + 3$ is also a multiple of 3. and so on. If $a$ is a multiple of 14, then $a + 14$ is also a multiple of 14. So, we only need to set $a$ to be a multiple of $2, 3, \ldots, 14$, then $a+2, a+3, \ldots, a+14$ are multiples of $2, 3, \ldots, 14$, respectively.

So, we set

$$a = 2 \times 3 \times 4 \times \cdots \times 14,$$

and the 13 consecutive natural numbers $a + 2, a + 3, \ldots, a + 14$ are all composite numbers.

**Example 8.** Given any natural number $N$, write out the digits of $N$ in reverse order to get a new natural number $N'$. Prove that $|N - N'|$ is divisible by 9.

**Solution** The remainder of $N$ divided by 9 is the same as the remainder of the digit sum of $N$ divided by 9. The remainder of $N'$ divided by 9 is the same as the remainder of the digit sum of $N'$ divided by 9. As $N$ and $N'$ have the same digits, but in reverse order, the digit sums of $N$ and $N'$ are the same. The remainder of $N$ divided by 9 is the same as the remainder of $N'$ divided by 9. So, $|N - N'|$ is divisible by 9.

## Reading

### The amazing Ramanujan

The Indian Ramanujan (1887–1920) can be said to be the most intuitive and insightful mathematician the world has seen.

On one occasion, the British mathematician Hardy (1877–1947) visited him in the hospital and said that the number 1729 of the taxi he took was an ordinary and boring number.

"No!" said Ramanujan, "it is the smallest number that can be written in two ways as the sum of two cubic numbers."

Indeed,

$$1729 = 10^3 + 9^3 = 12^3 + 1^3,$$

and any natural number smaller than 1729 cannot be written as the sum of two cubic numbers in two ways.

## Exercises

1. Find the number of positive integers within 200 that are divisible by 3, 4, and 5 at the same time.

2. Find the number of positive integers within 1000 that are divisible by 2, 4, 6, and 8 at the same time.

3. A six-digit number $\overline{3a123b}$ is divisible by 88. Find the values of $a$ and $b$.

4. Find the largest four-digit number that is divisible by 11 and 13.

5. For which numbers $x$ and $y$, the four-digit number $\overline{72xy}$ is divisible by 2, 3, 4, 5, 6, and 9?

6. Add three digits after 1992 to form a seven-digit number so that it is divisible by 2, 3, 5, and 11. Find the smallest one among such seven-digit numbers.

7. Write 17 consecutive natural numbers that are composite numbers.

8. Bob bought three fountain pens, five ballpoint pens, eight pencils, and twelve erasers, which cost 20.1 yuan. If each fountain pen costs 4 yuan, each ballpoint pen costs 0.8 yuan, and each eraser costs 0.05 yuan, did the salesperson compute correctly?

9. There is a three-digit number. If subtracted by 7, then it is divisible by 7. If subtracted by 8, then it is divisible by 8. If subtracted by 9, then it is divisible by 9. Find this three-digit number.

10. Find a two-digit number such that the remainder of this number divided by 3 is 1, the remainder of this number divided by 4 is 1, and the remainder of this number divided by 5 is also 1.

11. Find the number of positive integers within 100 that are neither divisible by 3 nor divisible by 4.

12. Choose four numbers from 5, 6, 7, 8, and 9 to form a four-digit number, such that this four-digit number is divisible by 3, 5, and 7. Among all these four-digit numbers that meet the conditions, what is the largest one?

13. If the remainders of 92, 118, and 157 divided by $n(n \neq 1)$ are the same, find the value of $n$.

14. Try to prove that, when we write the sum $1 + \frac{1}{2} + \frac{1}{3} + \frac{1}{4} + \cdots + \frac{1}{40}$ as a simplest fraction $\frac{m}{n}$, $m$ is not a multiple of 5.

# Chapter 20

# Odd Numbers and Even Numbers

Integers can be divided into two categories: odd numbers and even numbers. Integers that are not divisible by 2 are called odd numbers, and those that are divisible by 2 are called even numbers. The parity of integers has the following basic properties:

(1) An odd number cannot be equal to an even number.
(2) Even ± Even = Even;
   Even ± Odd = Odd;
   Odd ± Even = Odd;
   Odd ± Odd = Even.
   It is not difficult to find that during a calculation involving only integer additions and subtractions, if the number of odd numbers is even, then the result is even; if the number of odd numbers is odd, then the result is odd.
(3) Even × Even = Even;
   Even × Odd = Even;
   Odd × Odd = Odd.
   That is, the product of an odd number and an odd number is an odd number and the product of an odd number and an even number is an even number.
(4) An even number can be represented as $2k$, and an odd number can be represented as $2k + 1$ (or $2k - 1$), where $k$ is an integer.

Using the basic properties of parity, especially the simple property that odd numbers cannot be equal to even numbers, many mathematical problems can be solved.

**Example 1.** A small ferry boat travels between the left and right banks of a small river.

(1) The boat is on the left bank at the beginning, and after crossing the river several times, it returns to the left bank. Is the number of times the boat crosses the river odd or even? If it finally reaches the right bank, what happens?

(2) If the boat is on the left bank at the beginning, after crossing the river 99 times, which bank will it stop on, the left or the right?

**Solution**

(1) The boat is on the left bank at the beginning, crosses the river once to the right bank, crosses the river again, and returns from the right bank to the left bank. That is, every time it starts from the left bank to cross to the right bank and then returns to the left bank, it crosses the river twice. Therefore, if the boat starts from the left bank, crosses the river many times, and then returns to the left bank, then the number of crossings must be a multiple of 2, which is an even number. In the same way, it is not difficult to conclude that if the boat stops on the right bank at the end, the number of times of crossing the river must be an odd number.

(2) In (1), we found that if the boat is on the left bank at the beginning, it will return to the left bank after crossing the river for an even number of times; it will stop on the right bank after crossing the river for an odd number of times. Now, the boat crosses the river 99 times, which is an odd number of times. Therefore, at last, the boat stops on the right bank.

**Example 2.** Can $99^{99}$ and 99! (Remark: $99! = 1 \times 2 \times 3 \times 4 \times \cdots \times 99$; read as 99 factorial) be represented as sums of 99 consecutive odd numbers?

**Analysis**  $99^{99} = 99 \times 99^{98}$ is odd. Write $99^{98}$ first, then write 49 consecutive odd numbers after $99^{98}$, and write 49 consecutive odd numbers before $99^{98}$. The sum of these 99 consecutive odd numbers is exactly $99 \times 99^{98} = 99^{99}$.

On the other hand, 99! is even, and the sum of 99 odd numbers is odd.

## Solution

(1) $99^{99}$ can be represented as the sum of 99 consecutive odd numbers:

$$99^{99} = (99^{98} - 98) + (99^{98} - 96) + \cdots + (99^{98} - 2) + 99^{98}$$
$$+ (99^{98} + 2) + \cdots + (99^{98} + 96) + (99^{98} + 98).$$

It can be easily verified since the right-hand side is the sum of an arithmetic sequence.

(2) 99! cannot be represented as the sum of 99 consecutive odd numbers. As $99! = 1 \times 2 \times 3 \times \cdots \times 99$ is even and the sum of 99 consecutive odd numbers is odd, 99! cannot be represented as the sum of 99 consecutive odd numbers.

Remark: If the answer is yes, we often give concrete examples that satisfy the conditions of the problem, which is called the construction method.

If the answer is no, the method of proof by contradiction is often used.

**Example 3.** As shown in Figure 20.1, there is a house design in which each room has a door to the adjacent room. Bob is in a certain room, and he wants to go from this room to every room without repetition. Can it be done? If it can be done, which room should he be in at the beginning? And how should he go? If it cannot be done, please explain the reason.

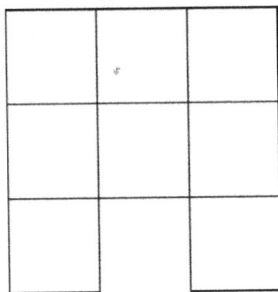

Fig. 20.1

**Solution**   It cannot be done.

Color the rooms in Figure 20.1, alternating with black and white as in Figure 20.2. In this way, no matter which room Bob starts from, he always goes from a white room to a black room or from a black room to a white room. Therefore, the move sequence must be white → black → white → black → $\cdots$ or black → white → black → white → $\cdots$. No matter which

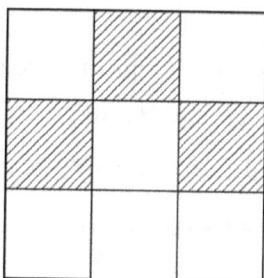

Fig. 20.2

way, the number of black rooms is equal to or different from the number of white rooms by 1. In Figure 20.2, there are 5 white rooms and 3 black rooms, so the difference is 2. Therefore, it is impossible to go through every room without repetition.

Remark:   Just as integers can be divided into odd and even numbers, we paint the rooms alternatingly in black and white to divide them into two categories. Several consecutive integers must be alternately odd and even, and the difference between numbers of odd and even numbers is at most one. Similarly, the sequence of the room colors is also alternately black and white. Therefore, the numbers of black and white rooms differ by at most one. This is the key to our solution of this example. Therefore, in essence, we still use parity to solve this problem. In fact, the problem can also be solved if, instead of painting the rooms in black and white, we paste the words "odd and even" in the rooms alternately.

**Example 4.** Paint the circles in Figure 20.3 arbitrarily in red or blue. Is it possible to make the number of red circles on each straight line an odd number? Please explain the reason.

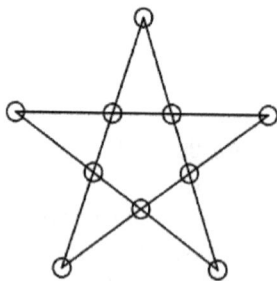

Fig. 20.3

**Solution**  If there is an odd number of red circles on each line, then the sum of the numbers of red circles on the 5 lines is still an odd number.

But on the other hand, when adding the number of red circles on the 5 lines, each circle is on two straight lines, so it is computed twice; then, the total sum should be an even number.

The results of the two aspects are contradictory. Therefore, it is impossible to make the number of red circles on each line odd.

**Example 5.** There are 19 × 19 intersection points on a Go board, and the intersection points are filled with black and white pieces alternately; that is, each white (or black) piece is surrounded above, below, left, and right by black (or white) pieces. Can you move all the blacks to the original positions of the whites and move all the whites to the original positions of the blacks?

**Solution**  It cannot be done.

As 19 × 19 = 361 is an odd number, there must be an odd number of white pieces and an even number of black pieces or an odd number of black pieces and an even number of white pieces. That is, the numbers of black and white pieces must be one odd and one even. An odd number cannot be equal to an even number, so it cannot be done.

**Example 6.** Many people who participated in a meeting shook hands with each other. Counting those who shook hands an odd number of times, is the number odd or even? Why?

**Solution**  For every handshake, each of the two people shakes hands once. Therefore, the sum of the numbers of handshakes of every person is twice the total number of handshakes. Therefore, the total number of handshakes among all participants must be an even number.

We divide the people participating in the meeting into two categories. The number of handshakes of type A is even, while the number of handshakes of type B is odd. The total number of handshakes of type A is obviously even. Note that the total number of handshakes of type A plus the total number of handshakes of type B is equal to the total number of handshakes of all participants, so the total number of handshakes of people of type B should also be an even number. The number of handshakes of each person of type B is an odd number and only an even number of odd numbers can have the sum as an even number; therefore, there must be an even number of people of type B; that is, the number of people who shook hands an odd number of times is an even number.

**Example 7.** Suppose that seven lights marked with $A$, $B$, $C$, $D$, $E$, $F$, and $G$ are arranged in a row in sequence, and each light is equipped with a switch. Now, the four lights $A$, $C$, $E$, and $G$ are on, and the rest are off. Starting with light $A$, Bob pulls the switches in sequence: from $A$ to $G$, then from $A$ to $G$, and so on. After pulling the switches 1999 times in this way, which lights are on?

**Solution**    After the switch of a light is pulled an odd number of times, the state changes, that is, on becomes off and off becomes on. After the switch of a light is pulled an even number of times, the state does not change, that is, on remains on and off remains off. Therefore, the key to this problem is to compute the parity of the number of times each switch is pulled. As

$$1999 = 7 \times 285 + 4,$$

we have that the switches of the four lights $A$, $B$, $C$, and $D$ have been pulled 286 times and the switches of the three lights $E$, $F$, and $G$ have been pulled 285 times. Therefore, the four lights marked $A$, $B$, $C$, and $D$ do not change state, and the three lights marked $E$, $F$, and $G$ change state. As the four lights $A$, $C$, $E$, and $G$ are on at the beginning, we infer that the three lights $A$, $C$, and $F$ are on at the end.

**Example 8.** There are seven cups on the table, all of which are upright. If four cups are flipped at a time, is it possible to make all the cups upside down after several such flips?

**Solution**    It is impossible.

We denote the cup upright as 0 and the cup upside down as 1. At the beginning, as all seven cups are upright, the sum of these seven numbers is 0, which is an even number.

Every time a cup is flipped, its number changes from 0 to 1 or from 1 to 0, which changes the parity. As four cups are flipped each time, the parity of the sum of these seven numbers changes four times, and thus the parity of the sum is still the same as before. Therefore, no matter how many times they are flipped, the sum of these seven numbers remains the same, and it is still an even number.

When the cups are all upside down, the sum of these seven numbers is 7, which is an odd number. Therefore, no matter how many times the cups are flipped, it is impossible to turn all the cups upside down.

**Example 9.** Suppose that $a_1, a_2, \ldots, a_{2005}$ is an arbitrary arrangement of $1, 2, \ldots, 2005$. Try to prove that $(a_1 - 1)(a_2 - 2) \cdots (a_{2005} - 2005)$ must be an even number.

**Proof** The numbers $1, 2, \ldots, 2005$ have one more odd number than even numbers, so there is one more odd number than even numbers in $a_1$, $a_2, \ldots, a_{2005}$. In this way, among the differences $a_1 - 1, a_3 - 3, \ldots, a_{2005} - 2005$, there is at least one with the form of an odd number minus another odd number. This difference is even. Thus, $(a_1 - 1)(a_2 - 2) \cdots (a_{2005} - 2005)$ is even.

## Reading

---

### Goldbach Conjecture

Goldbach (1690–1764), a German minister in Petersburg, Russia, was an amateur mathematician. He wrote to his friend, the great mathematician Euler, in 1742, saying that he conjectured the following:

Every odd number greater than 7 can be expressed as the sum of three prime numbers.

Euler wrote back that Goldbach's conjecture can be deduced from the following conjecture:

Every even number greater than 4 is the sum of two prime numbers.

This is the famous Goldbach conjecture, or Goldbach–Euler conjecture. In fact, the French mathematician Descartes (1596–1650) had already proposed this conjecture.

For small even numbers, Goldbach's conjecture is not difficult to verify. For example,

$$6 = 3 + 3,$$
$$8 = 3 + 5,$$
$$10 = 3 + 7,$$
$$12 = 5 + 7,$$

$$\vdots$$

In fact, all even numbers below 300 million have been verified. But the number of even numbers is infinite. Even if you verify more numbers, there are still infinitely many numbers without verification. Therefore, the Goldbach conjecture cannot be solved by verification.

We know that each prime number has only one prime factor, namely itself. If a number has at most two prime factors, we record it as $p_2$, that is, $p_2 = p$ or $pq$, where $p$ and $q$ are prime numbers (which can be the same). Then, $p_2$ is an almost prime number.

Chinese mathematicians have made important contributions to the Goldbach problem. Chen Jingrun proved "1 + 2," that is, every sufficiently large even number is equal to the sum of a prime number and an almost prime number $p_2$. However, until now, no one has been able to prove "1+1," that is, to remove the "almost" of "almost prime number" and prove that every sufficiently large even number is equal to the sum of one prime number and another prime number.

---

**Exercises**

1. There is a class of students, each of whom participates in one of the three interest groups of Mathematics, Chinese, or English. The number of students participating in the Mathematics interest group is 3 more than the number of students participating in the Chinese interest group. The number of students participating in the Chinese interest group is 5 more than the number of students participating in the English interest group. The number of students participating in the English interest group is an even number. Is the number of students in this class odd or even?

2. Find the sum of the first 10 odd numbers in positive integers. Generally, find the sum of the first $n$ odd numbers in positive integers.

3. Tom crosses a few rivers. When he enters a river, he takes off his shoes; when he goes ashore, he puts on his shoes. Initially, Tom is in the river, and he has taken off or put on the shoes an even number of times. Is Tom now in the river or on the bank?

4. The sum of seven consecutive odd numbers is 399. Find these seven numbers.

5. The product of three consecutive even numbers is a six-digit number $\overline{8****2}$. Find these three even numbers.

6. If two people talk on the phone once, it is considered that these two people have called once. Is the number of people in the world who have an odd number of calls odd or even?

7. Five lights, $A$, $B$, $C$, $D$, and $E$, are all off at the beginning. Bob pulls the switches of these five lights in sequence; that is, he pulls the switches from $A$ to $E$, then pulls the switches from $A$ to $E$, and so on. After Bob pulls the switches 189 times, which lights are on?

8. There are four cups on the table, all of which are upright. Flip three cups at a time. Can all the cups be turned upside down after flipping several times?

9. Can you fill in a plus sign or a minus sign in each box of the following formula to make the equation true?

$$1\square2\square3\square4\square5\square6\square7\square8\square9 = 10.$$

10. Can you find natural numbers $a$ and $b$ such that $a^2 = 2002 + b^2$?

11. There are 9 cups, all of which are upright. Flip four of them at a time. Can all the cups be upside down after several such flips?

12. As shown in Figure 20.4, there are 9 rooms in a house, and each number indicates the number of the room. Each room has a door connected to an adjacent room. Bob needs to move from the first room, go through the nine rooms without repeating, and finally return to the first room. Can he make it?

| 1 | 2 | 3 |
|---|---|---|
| 4 | 5 | 6 |
| 7 | 8 | 9 |

Fig. 20.4

13. There are 9 points on a plane, and every three points are not on the same straight line. We want to draw exactly three straight lines from each point and connect them with any three of the remaining points. Can this idea be realized?

14. There is a knight on a chess board, and after jumping a few steps, it returns to its original position. Is the number of jumps odd or even? Why?

15. Among the three numbers $a$, $b$, and $c$, one is 2001, one is 2002, and the last one is 2003. Try to prove that the product of $(a+1)$, $(b+2)$, and $(c+3)$ must be an even number.

# Chapter 21

# Prime Numbers and Composite Numbers

For an integer greater than 1, if it has only two divisors, that is, 1 and itself, we call it a prime number. If there are more than two divisors, that is, a divisor different from 1 and itself, we call it a composite number. Positive integers can be divided into three categories: prime numbers, composite numbers, and 1, that is,

$$\text{Positive integers} \begin{cases} \text{prime numbers,} \\ \text{composite numbers,} \\ 1. \end{cases}$$

It should be noted that 1 is neither a prime nor a composite number. We call it a unit.

Prime numbers and composite numbers have the following common properties:

(1) There are infinitely many prime numbers.

(2) 2 is the only integer that is both a prime number and an even number, that is, the only even prime number. A prime number greater than 2 must be an odd number.

(3) If the prime number $p \mid ab$, then there must be $p \mid a$ or $p \mid b$.

(4) If the product of positive integers $a$ and $b$ is a prime number $p$, then it must be $a = p$ or $b = p$.

(5) Unique factorization theorem: any integer $n$ $(n > 1)$ can be uniquely factorized into $n = p_1^{a_1} p_2^{a_2} \cdots p_k^{a_k}$, where $p_1 < p_2 < \cdots < p_k$ are prime numbers and $a_1, a_2, \ldots, a_k$ are positive integers.

**Example 1.** There are four numbers: one is the smallest odd prime number, one is an even prime number, one is the largest prime number less than 30, and the last one is the smallest prime number greater than 70. Find their sum.

**Solution**   The smallest odd prime number is 3, the only even prime number is 2, the largest prime number less than 30 is 29, and the smallest prime number greater than 70 is 71. Therefore, their sum is

$$3 + 2 + 29 + 71 = 105.$$

Remark:   When solving problems related to prime numbers, it is helpful to use some commonly known properties, such as that 1 is neither a prime number nor a composite number, 2 is the only even prime number and the smallest prime number, and 3 is the smallest odd prime number. In addition, within 100, there are 25 prime numbers in total; they are 2, 3, 5, 7, 11, 13, 17, 19, 23, 29, 31, 37, 41, 43, 47, 53, 59, 61, 67, 71, 73 , 79, 83, 89, and 97.

**Example 2.** The sum of seven different prime numbers is 60. Find the smallest one among them.

**Solution**   If seven different prime numbers are all odd prime numbers, then their sum must be an odd number and cannot be equal to 60, so there are even numbers among these seven different prime numbers. We know that 2 is the only even prime number, so 2 must be among these seven prime numbers. 2 is the smallest one of all prime numbers, so the smallest prime number among these seven prime numbers is 2.

Remark:   We use the characteristic that 2 is the only even prime number and the smallest prime number in this problem. It is not difficult to conclude that these seven prime numbers are 2, 3, 5, 7, 11, 13, and 19.

**Example 3.** If $n$ is a positive integer and $n + 3$ and $n + 7$ are both prime numbers, then find the remainder of $n$ divided by 3.

**Solution**   As we know, the remainder of $n$ divided by 3 can only be 0, 1, or 2.

If the remainder is 0, that is, $n = 3k$ ($k$ is a positive integer), then $n + 3 = 3k + 3 = 3(k + 1)$, so $3 \mid n + 3$. Also, as $3 \neq n + 3$, so $n + 3$ is not a prime number, which contradicts the problem assumption.

If the remainder is 2, that is, $n = 3k + 2$ ($k$ is a non-negative integer), then $n + 7 = 3k + 2 + 7 = 3(k + 3)$, so $3 \mid n + 7$. Then, $n + 7$ is not a prime number, which contradicts the problem assumption.

Therefore, the remainder of $n$ divided by 3 can only be 1.

Remark: When an integer is divided by $m$, the remainder may be $0, 1, \ldots$, or $m-1$, totaling $m$ types. All integers can be classified according to the remainder divided by $m$ into $m$ categories. For example, when $m = 2$, the remainder can only be 0 or 1, so it can be divided into two categories: one is an integer with a remainder of 0 when divided by 2, that is, an even number, and the other is an integer with a remainder of 1 when divided by 2, that is, an odd number. Similarly, when $m = 3$, integers can be divided into three categories. That is, when divided by 3, the remainders are 0, 1, or 2. Classifying whether the remainders are the same is an important method in number theory, which has a wide range of applications.

**Example 4.** Suppose $n_1$ and $n_2$ are two arbitrary prime numbers greater than 3. Let $N_1 = n_1^2 - 1$ and $N_2 = n_2^2 - 1$. Find the minimum value of the greatest common divisor of $N_1$ and $N_2$.

**Solution** As $n_1$ is a prime number greater than 3, so $n_1$ is not a multiple of 3, $n_1 = 3k + 1$, or $3k + 2$.

When $n_1 = 3k + 1$, we have that $n_1 - 1 = 3k$ is a multiple of 3. When $n_1 = 3k + 2$, we have that $n_1 + 1 = 3k + 3$ is a multiple of 3. In either case,

$$N_1 = n_1^2 - 1 = (n_1 + 1)(n_1 - 1)$$

is a multiple of 3.

Also, as $n_1$ is an odd number, $n_1 = 4k + 1$ or $4k + 3$.

When $n_1 = 4k + 1$, we have that $n_1 + 1 = 4k + 2$ is a multiple of 2 and $n_1 - 1 = 4k$ is a multiple of 4, so $N_1$ is a multiple of 8. When $n_1 = 4k + 3$, we have that $N_1$ is, for the same reason, a multiple of 8.

As 3 and 8 are co-prime, $24 \mid N_1$.

Similarly, we have $24 \mid N_2$.

In addition, let $n_1 = 5$, then $N_1 = 24$. This indicates that the minimal greatest common divisor of $N_1$ and $N_2$ cannot be larger than 24.

In summary, the greatest common divisor of $N_1$ and $N_2$ is at least 24.

Remark: From the above example, we draw two useful conclusions:

(1) If $n$ is not a multiple of 3, then the remainder of $n^2$ divided by 3 is 1.
(2) If $n$ is an odd number, then the remainder of $n^2$ divided by 8 is 1.

**Example 5.** Someone said, "There must be a prime number among any seven consecutive integers." Is it correct?

**Solution** It is incorrect. For example, consider 90, 91, 92, 93, 94, 95, and 96; these seven consecutive integers are all composite numbers, not prime numbers.

**Remark:**   See Example 7 in Chapter 19.

**Example 6.** Suppose that natural numbers $n_1 > n_2$ and $n_1^2 - n_2^2 = 79$. Try to find the values of $n_1$ and $n_2$.

**Solution**   Based on the problem assumption, we have $n_1^2 - n_2^2 = (n_1 + n_2)(n_1 - n_2) = 79$. As $n_1 > n_2$, $n_1 + n_2$ and $n_1 - n_2$ are both positive integers. But 79 is a prime number, according to the property of prime numbers, and $n_1 + n_2 > n_1 - n_2$, so we have

$$\begin{cases} n_1 + n_2 = 79, \\ n_1 - n_2 = 1. \end{cases}$$

It is not difficult to find that $n_1 = 40$ and $n_2 = 39$.

**Example 7.** Suppose that $n$ is an even number not less than 40. Try to prove that $n$ can always be represented as the sum of two odd composite numbers.

**Proof**   As $n$ is an even number, so the ones digit of $n$ must be one of 0, 2, 4, 6, and 8.

(1) If the ones digit of $n$ is 0, then $n = 15 + 5k$ ($k \geqslant 5$ is an odd number).
(2) If the ones digit of $n$ is 2, then $n = 27 + 5k$ ($k \geqslant 3$ is an odd number).
(3) If the ones digit of $n$ is 4, then $n = 9 + 5k$ ($k \geqslant 7$ is an odd number).
(4) If the ones digit of $n$ is 6, then $n = 21 + 5k$ ($k \geqslant 5$ is an odd number).
(5) If the ones digit of $n$ is 8, then $n = 33 + 5k$ ($k \geqslant 3$ is an odd number).

To conclude, any even number not less than 40 can be represented as the sum of two odd composite numbers.

**Example 8.** Prove that there are infinitely many $n$, such that the polynomial $n^2 + 3n + 7$

(1) represents a composite number;
(2) is a multiple of 11.

**Proof**   We only need to prove (2).
When $n = 11k + 1 (k \geqslant 1)$,

$$n^2 + 3n + 7 = (11k + 1)^2 + 3(11k + 1) + 7$$
$$= 11(11k^2 + 5k + 1)$$

is a multiple of 11. Since $11k^2 + 5k + 1 > 1$, $n^2 + 3n + 7$ is a composite number.

## Reading

---

### $\sqrt{2}$ is an Irrational Number

From the reading in Chapter 2, we know that for a unit square with a side length of 1, if a square is drawn with its diagonal as the side, the area of the new square is twice that of the unit square, that is, the area is 2. So, the square of the side length of the new square is 2.

The number whose square is 2 is given by the square root of 2. There are two such numbers. The positive one is written as $\sqrt{2}$, and the negative one is the opposite number of $\sqrt{2}$ (obviously, the square of $-\sqrt{2}$ is also 2).

From Figure 2.3, it is easy to see that $\sqrt{2} > 1$ and $\sqrt{2} < 2$. In fact,

$$\sqrt{2} = 1.41421356\ldots$$

is an irrational number.

How do we prove that $\sqrt{2}$ is an irrational number, that is, it is not a rational number and cannot be expressed in the form of $\frac{m}{n}$ (where $m$ and $n$ are positive integers)?

Suppose there is

$$\sqrt{2} = \frac{m}{n}, \tag{21.1}$$

where $m$ and $n$ are co-prime positive integers. Squaring both sides and multiplying by $n^2$, we get

$$2n^2 = m^2. \tag{21.2}$$

Decompose both sides of (21.2) into factors, that is, write $m$ and $n$ as the product of prime factors. Since the right side of (21.2) is the square number $m^2$, the exponent of the prime factor 2 on the right side should be an even number (including 0). Similarly, in the decomposition formula of $n^2$, the exponent of 2 should also be an even number; however, in the decomposition formula of $2n^2$, the exponent of 2 is an odd number (an even number plus 1 is an odd number). Therefore, the exponents of 2 on both sides of (21.2) are not equal. (21.2) cannot be established, that is, $\sqrt{2}$ is an irrational number.

Similarly, it can be proved that $\sqrt{3}$, $\sqrt{5}$, $\sqrt{6}$, and $\sqrt{7}$ as well as, in general, $\sqrt{k}$ (where $k$ is not a square number) are all irrational numbers.

$\pi$ is also an irrational number. But proving it is difficult.

---

**Exercises**

1. There are three positive integers: one is the smallest odd prime number, one is the smallest odd composite number, and the last one is neither a prime nor a composite number. Find the product of these three numbers.

2. There are three numbers: one is an even prime number, one is the smallest prime number greater than 50, and the last one is the largest prime number within 100. Find the sum of these three numbers.

3. The sum of two prime numbers is 49. Find the product of these two prime numbers.

4. Suppose that $p_1$ and $p_2$ are two prime numbers greater than 2. Prove that $p_1 + p_2$ is a composite number.

5. Suppose that $p$ is a prime number and $p^2 + 3$ is also a prime number. Prove that $p^3 + 3$ is a prime number.

6. Suppose that $p$ ($p > 3$) and $p+2$ are prime numbers. Find the remainder of $p$ divided by 3.

7. If natural numbers $n_1 > n_2$ and $n_1^2 - n_2^2 - 2n_1 - 2n_2 = 19$, then find the values of $n_1$ and $n_2$.

8. What is the smallest positive integer that has four different prime factors?

9. Find the sum of all the different prime factors of 2000.

10. Try to prove that positive integers in the form $111111 + 9 \times 10^k$ (where $k$ is a non-negative integer) must be composite.

11. If $n$ is a positive integer and $n + 3$ and $n + 7$ are prime numbers. Find the remainder of $n$ divided by 6.

12. Suppose that $n$ is a natural number. Try to prove that $10 \mid n^5 - n$.

13. Prove that there are infinitely many $n$, such that $n^2 + n + 41$

   (1) represents a composite number;
   (2) is a multiple of 43.

14. Try to prove that there are infinitely many prime numbers among natural numbers.

15. There are nine consecutive natural numbers, each of which is greater than 80. At most, how many prime numbers are among them?

# Chapter 22

# The Rule of Sum and the Rule of Product

With the rapid development of computer science, combinatorics, an ancient subject of mathematics, has shown new vitality. Therefore, related problems are often encountered in mathematics competitions. This chapter only discusses the two most basic combinatorial rules, which are the rule of sum and the rule of product. These two rules can be described as follows:

**The rule of sum:** We are going to do one thing, and there are $n$ things to choose from. There are $m_1$ different ways to do the first thing, $m_2$ different ways to do the second thing, ... and $m_n$ different ways to do the $n$th thing. Then, the total number of different ways to complete it is

$$m_1 + m_2 + \cdots + m_n.$$

**The rule of product:** To do one thing, it can be divided into $n$ steps: there are $m_1$ different ways to do the first step, $m_2$ different ways to do the second step, ..., $m_n$ different ways to do the $n$th step. Then, there are $m_1 \cdot m_2 \cdots m_n$ different ways to complete it.

**Example 1.** From $A$ to $B$, one can take a train, a car, or a ship. There are 4 trains, 8 cars, and 3 ships. How many different ways are there to go from $A$ to $B$?

**Solution** There are three means of transportation to go from $A$ to $B$. The first is by train, in which there are 4 different ways. The second is by car, in which there are 8 different ways. The third is by boat, in which there are 3 different ways. According to the rule of sum, there are $4 + 8 + 3 = 15$ different ways to go from $A$ to $B$.

**Example 2.** There are 5 balls in one box and 6 balls in the other box, and these balls are all different.

(1) How many different ways are there to pick a ball from these two boxes?
(2) How many different ways are there to take a ball from each of the two boxes?

**Solution**

(1) Take a ball from one of the two boxes. There are two methods. The first method is to take the ball from the first box. There are 5 different ways. The second method is to take the ball from the second box. There are 6 different ways. According to the rule of sum, there are $5 + 6 = 11$ different ways to take the ball.

(2) Take a ball from each of the two boxes, which can be divided into two steps. The first step is to take a ball from the first box, in which there are 5 different ways. The second step is to take a ball from the second box, in which there are 6 different ways. After the two steps are completed, this thing (taking a ball from each box) is completed. According to the rule of product, there are $5 \times 6 = 30$ different ways to take the balls.

Remark: From the above example, we can see the difference: when using the rule of sum, there are $n$ types of methods to complete one thing, and no matter which type of method is used, the thing can be completed. However, when using the rule of product, the thing is divided into $n$ steps, but no matter which step is completed, it is only a part of the thing. Only when all steps are completed is the thing completed. Therefore, any one of the $n$ steps is indispensable. This is the main difference between the rule of product and the rule of sum.

**Example 3.** There are three groups, $A$, $B$, and $C$. Group $A$ has 6 people, group $B$ has 5 people, and group $C$ has 4 people. Choose one person from each group to attend the meeting together. How many ways are there? If the three groups jointly elect a representative, how many ways are there?

**Solution**

(1) If each group elects one person to attend the meeting together, then the election in each group is only one step to complete the entire election. After using the rule of product, we have that there are $6 \times 5 \times 4 = 120$ ways.

(2) If the three groups jointly elect a representative, then any group elects a representative, and the entire election has been completed. After using the rule of sum, there are $6 + 5 + 4 = 15$ ways.

**Example 4.** Multiply three polynomials

$$a_l x^l + a_{l-1} x^{l-1} + \cdots + a_1 x + a_0,$$

$$b_m x^m + b_{m-1} x^{m-1} + \cdots + b_1 x + b_0,$$

$$c_n x^n + c_{n-1} x^{n-1} + \cdots + c_1 x + c_0$$

$(a_l, b_m, c_n \neq 0)$ together. Before merging similar terms, how many terms are there at most? After merging similar terms, how many terms are there at most?

**Solution** The three polynomials have at most $l + 1$, $m + 1$, and $n + 1$ terms, respectively (some of which may have coefficients of 0). When multiplying, one term from each of the three polynomials is multiplied, and then the similar terms are merged. So, according to the rule of product, before merging similar terms, the product has at most $(l+1)(m+1)(n+1)$ terms (if $a_i$, $b_j$, and $c_k$ are all nonzero, $0 \leqslant i \leqslant l$, $0 \leqslant j \leqslant m$, $0 \leqslant k \leqslant n$, then there are exactly $(l + 1)(m + 1)(n + 1)$ terms).

The degree of the highest order term of the product is $l + m + n$, so after merging similar terms, there are at most $l + m + n + 1$ terms.

**Example 5.** Figure 22.1 shows a $5 \times 5$ square. Put five chess pieces $A$, $B$, $C$, $D$, and $E$ in the squares such that only one chess piece appears in each row and each column. How many ways are there in total?

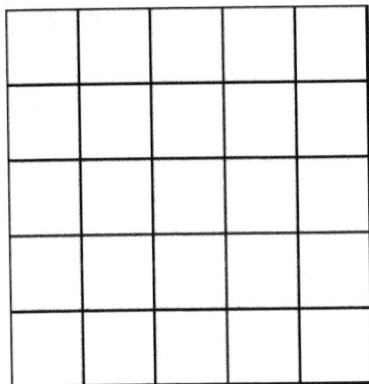

Fig. 22.1

**Solution**   When putting the five chess pieces $A$, $B$, $C$, $D$, and $E$ in a $5 \times 5$ square, it can be divided into five steps. The first step is to put $A$, and it can be placed in any of the $5 \times 5 = 25$ squares. So, there are 25 ways to put it. In the second step, put $B$; according to the problem restriction, $B$ can be neither in the same row as $A$, nor in the same column as $A$. After $A$ occupies a square, all other squares in the same row and the same column of this square cannot hold $B$ any longer. Therefore, $B$ has $4 \times 4 = 16$ squares to choose from (eliminating the row and column occupied by $A$ and leaving $4 \times 4$ squares). Similarly, in the third step, $C$ cannot be put in the same row or column as $A$ or $B$, so there are only $3 \times 3 = 9$ ways to put it. In the fourth step, $D$ cannot be put in the same row or column as $A$, $B$, or $C$, so there are only $2 \times 2 = 4$ ways to put it. In the fifth step, $E$ cannot be put in the same row or column as $A$, $B$, $C$, or $D$, so there is only $1 \times 1 = 1$ way to put it. According to the rule of product, the total number of ways to put them in the $5 \times 5$ square is

$$25 \times 16 \times 9 \times 4 \times 1 = 14400.$$

**Example 6.** Figure 22.2 is a map of seven counties under the jurisdiction of a certain city. Color the map with five colors of red, green, blue, purple, and black, such that any two adjacent counties have different colors. How many different ways are there?

**Solution**   For convenience, under the premise of keeping the adjacent relation of each county unchanged, we change Figure 22.2 to Figure 22.3.

Coloring is done in the order of $A$, $B$, $C$, $D$, $E$, $F$, and $G$. $A$ has 5 ways. $B$ cannot have the same color as $A$, so there are 4 ways. $C$ cannot have the same color as $A$ or $B$, so there are 3 ways. $D$ cannot have the

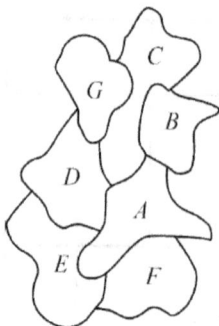

Fig. 22.2

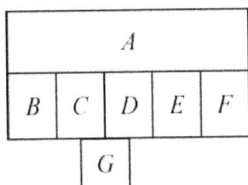

Fig. 22.3

same color as $A$ or $C$, so there are 3 ways. $E$ cannot have the same color as $A$ or $D$, so there are 3 ways. $F$ cannot have the same color as $A$ or $E$, so there are 3 ways. Finally, $G$ cannot have the same color as $C$ or $D$, and thus there are 3 ways.

According to the rule of product, the map has $5 \times 4 \times 3 \times 3 \times 3 \times 3 \times 3 = 4860$ different ways of coloring.

**Example 7.** Consider positive integers whose digits are all from 0, 1, 2, 3, 4, 5, and 6:

(1) How many four-digit numbers are there? Among them, how many odd and even numbers are there?
(2) How many four-digit numbers without repeated digits are there? Among them, how many odd and even numbers are there?

**Solution**

(1) Let's consider the digits one by one.

The first digit of this four-digit number cannot be 0, and there are 6 choices (that is, choosing any number from 1 to 6). For the remaining digits, each digit can be chosen from the 7 numbers from 0 to 6, and there are 7 different ways. So there are $6 \times 7 \times 7 \times 7 = 2058$ four-digit numbers.

Among these four-digit numbers, the number of odd numbers can be obtained in a similar way.

The first digit cannot be 0, so there are 6 ways. The hundreds digit and tens digit can be selected from 0 to 6, each in 7 ways. The ones digit can only be 1, 3, or 5, so there are 3 ways. Therefore, the number of odd numbers is $6 \times 7 \times 7 \times 3 = 882$.

Among these four-digit numbers, the number of even numbers can be obtained by the following two methods.

Method one is exactly the same as the method for counting odd numbers. It is not difficult to obtain the number of even numbers as $6 \times 7 \times 7 \times 4 = 1176$.

In method two, the number of even numbers is equal to the number of four-digit numbers minus the number of odd numbers, that is, $2058 - 882 = 1176$.

(2) Use the same method to find how many four-digit numbers there are without repeated digits.

The first digit cannot be 0, so there are only 6 ways. The hundreds digit cannot be the same as the first digit, and there are 6 ways when the first digit is removed from the 7 numbers from 0 to 6. The tens digit cannot be the same as the first two digits, so there are 5 ways. The ones digit cannot be the same as the first three digit, so there are 4 ways. Therefore, there are $6 \times 6 \times 5 \times 4 = 720$ four-digit numbers without repeated digits.

Among them, the number of odd numbers can be obtained by the following method: the ones digit can only be 1, 3, or 5, so there are 3 ways. The first digit cannot be 0, nor can it be the same as the ones digit, thus there are 5 ways. The hundreds digit cannot be the same as the ones digit or the first digit, so there are 5 ways. The tens digit cannot be the same as the ones digit, the first digit, or the hundreds digit, so there are 4 ways. Therefore, the number of odd numbers is $3 \times 5 \times 5 \times 4 = 300$.

The number of even numbers is $720 - 300 = 420$.

Remark: To find the number of even numbers without repeated digits, the following method can also be used: such even numbers can be divided into two types. One type is that its ones digit is 0, then there are 6 choices for the thousands digit, 5 choices for the hundreds digit, and 4 choices for the tens digit. Therefore, there are $6 \times 5 \times 4 = 120$ numbers of this type. Another type is that its ones digit is 2, 4, or 6, and there are 3 ways. Now, there are 5 choices for the thousands digit (which cannot be 0 and cannot be the same as the ones digit), the hundreds digit has 5 choices (not the same as the ones and thousands digits), and the tens digit has 4 choices. So, there are $3 \times 5 \times 5 \times 4 = 300$ numbers of this type. According to the rule of sum, there are $120 + 300 = 420$ even numbers. The result is the same as before. This solution comprehensively applies the rule of sum and the rule of product.

Consider the following carefully: why do we consider the two types separately that the ones digit is 0 and the ones digit is not 0?

**Example 8.** Arbitrarily divide 10 people into two groups, A and B, with at least one person in each group. How many different ways of dividing are there?

**Solution** As everyone can be assigned to either group A or group B, there are two ways to assign each person. Therefore, according to the rule of product, 10 people can be divided in

$$\underbrace{2 \times 2 \times \cdots \times 2}_{10} = 2^{10} = 1024$$

different ways.

But among these 1024 divisions, one is that all 10 people are in group A and there is no one in group B; another one is that all 10 people are in group B and there is no one in group A. These two divisions do not meet the condition of this problem. So, there are a total of $1024 - 2 = 1022$ different divisions that meet the condition of this problem.

Remark: Sometimes, we first find the total number of ways and then remove those that do not meet the condition of the problem. This is usually an efficient method.

**Exercises**

1. There are three rows of books on a bookshelf. There are 12 books in the first row, 20 books in the second row, and 15 books in the third row. Bob wants to choose a book to read from them. How many different ways does he have?

2. There are 18 boys and 15 girls in a class. One of them is selected to participate in the summer camp. How many different ways of selecting are there?

3. There are 2 balls in the first pocket, 4 balls in the second pocket, and 5 balls in the third pocket. The balls are all different.

   (1) How many different ways are there to take a ball from the pockets?
   (2) How many different ways are there to draw a ball from each of the three pockets?

4. As shown in Figure 22.4, there are two pathways from point $A$ to point $B$, three pathways from point $B$ to point $C$, and four pathways

from point $A$ to point $C$. How many different ways are there to go from point $A$ to point $C$?

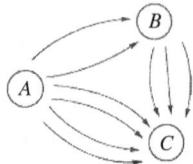

Fig. 22.4

5. In the expansion of the polynomial

$$(a_1 + a_2 + a_3)(b_1 + b_2 + b_3 + b_4)(c_1 + c_2),$$

how many different terms are there?

6. Find the number of positive divisors of 2000.

7. How many different three-digit numbers can be formed with the four digits 1, 2, 3, and 4 with repetitions allowed?

8. Divide 6 people into two groups, $A$ and $B$, with at least one person in each group. How many different ways are there?

9. An express train from Nanjing to Shanghai has to stop at six major stations on the way. How many different kinds of tickets will the railway bureau prepare for this express train? At most, how many different fares are there among these tickets?

10. How many different arrangements are there to let 4 people stand in a row to take a group photo?

11. Use the three digits 2, 3, and 4 to form a three-digit number without repeated digits.

    (1) Find the sum of the digit sums of these three-digit numbers.
    (2) Find the sum of these three-digit numbers.

12. Among all positive divisors of 2000, how many even numbers are there?

13. How many numbers can be formed with the digits 0, 1, 2, 3, and 4:

    (1) four-digit numbers?
    (2) four-digit even numbers?
    (3) four-digit numbers without repeated digits?
    (4) four-digit even numbers without repeated digits?
    (5) positive integers without repeated digits?

14. Drop three letters $A$, $B$, and $C$ randomly into four different mailboxes. How many different ways are there to drop the letters?

15. There are 5 people standing in a row to take pictures, and one of them must stand in the middle. How many ways can they stand?
16. How many three-digit numbers are divisible by 3 and contain the number 9?
17. As shown in Figure 22.5, a map with five parts $A$, $B$, $C$, $D$, and $E$ is to be colored with four different colors. Adjacent parts cannot have the same color, and non-adjacent parts can have the same color. How many different ways of coloring are there?

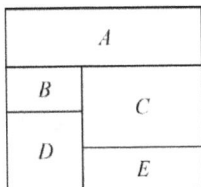

Fig. 22.5

# Chapter 23

# Number of Divisors

We know that if $a$ and $b$ are both integers and there is an integer $c$ such that

$$a = bc, \tag{23.1}$$

then $b$ is called a divisor of $a$.

Usually, we only discuss positive divisors of positive integers, that is, $a$, $b$, and $c$ in (23.1) are all positive integers. Unless otherwise stated, all letters represent positive integers.

How many divisors does 72 have?

It is not difficult to list them one by one. There are 12 divisors of 72. They are

$$1, 2, 3, 4, 6, 8, 9, 12, 18, 24, 36, 72. \tag{23.2}$$

Note that this includes 1 and 72 itself.

Is there a formula that can help us compute the number of divisors?

Yes, there is.

If we decompose 72 into a product of prime factors, we get

$$72 = 2^3 \times 3^2. \tag{23.3}$$

Then, all divisors of 72 are in the form

$$b = 2^{\beta_1} \times 3^{\beta_2}, \tag{23.4}$$

where $\beta_1$ can take four values: 0, 1, 2, and 3, while $\beta_2$ can take three values: 0, 1, and 2. (For example: when $\beta_1 = 0$ and $\beta_2 = 0$, $b = 1$; when $\beta_1 = 3$ and $\beta_2 = 2$, $b = 72$.)

Therefore, the number of divisors of 72 is

$$4 \times 3 = 12.$$

In general, suppose a natural number $n$ can be decomposed into

$$n = p_1^{\alpha_1} p_2^{\alpha_2} \cdots p_k^{\alpha_k}, \tag{23.5}$$

where $p_1, p_2, \ldots, p_k$ are different prime numbers and $\alpha_1, \alpha_2, \ldots, \alpha_k$ are positive integers, then all divisors of $n$ have the form

$$p_1^{\beta_1} p_2^{\beta_2} \cdots p_k^{\beta_k}, \tag{23.6}$$

where $\beta_1$ can take $\alpha_1 + 1$ values: $0, 1, 2, \ldots, \alpha_1$; $\beta_2$ can take $\alpha_2 + 1$ values: $0, 1, 2, \ldots, \alpha_2; \ldots; \beta_k$ can take $\alpha_k + 1$ values: $0, 1, 2, \ldots, \alpha_k$. So, the total number of divisors of $n$ is

$$(\alpha_1 + 1)(\alpha_2 + 1) \cdots (\alpha_k + 1). \tag{23.7}$$

**Example 1.** Find the number of divisors of

(1) 6000,
(2) 2006.

**Analysis**    First, we decompose 6000 and 2006 into factors, and then we use the above formula (23.7).

**Solution**

(1) $6000 = 2 \times 3 \times 10^3 = 2^4 \times 3 \times 5^3$, so the number of divisors of 6000 is

$$(4+1) \times (1+1) \times (3+1) = 40.$$

(2) $2006 = 2 \times 17 \times 59$, so the number of divisors of 2006 is

$$(1+1) \times (1+1) \times (1+1) = 8.$$

**Example 2.** What is the smallest number that has exactly 10 divisors?

**Analysis**    The problem is the inverse of example 1; that is, given that

$$(\alpha_1 + 1)(\alpha_2 + 1) \cdots (\alpha_k + 1) = 10, \tag{23.8}$$

find the value of $n$.

In order to find the value of $n$, we should first determine the value of $k$. The value of $k$ can be 1 or 2.

**Solution** $10 = 2 \times 5$, so in (23.8), $k$ can be 1 or 2.

If $k$ is 1, then

$$\alpha_1 + 1 = 10, n = p_1^{\alpha_1}.$$

$$\alpha_1 = 9, n = p_1^9.$$

To make $n$ smallest, $p_1$ should be 2, $n = 2^9$.

If $k$ is 2, then

$$(\alpha_1 + 1)(\alpha_2 + 1) = 2 \times 5, n = p_1^{\alpha_1} p_2^{\alpha_2}.$$

So (without order), we have

$$\alpha_1 = 4, \alpha_2 = 1, n = p_1^4 p_2.$$

In order to make $n$ smallest, $p_1$ should be 2 and $p_2$ should be 3, $n = 2^4 \times 3$.

As $2^9 > 2^4 \times 3$, the smallest number satisfying the problem condition is $2^4 \times 3 = 48$.

**Example 3.** If $n$ is a square number, then prove that the number of divisors of $n$ must be odd.

**Analysis** Use the decomposition of $n$ and the formula (23.7). Or, try to pair the divisors of $n$.

**Solution** Decompose $n$ into a product of prime factors:

$$n = p_1^{\alpha_1} p_2^{\alpha_2} \cdots p_k^{\alpha_k}. \tag{23.9}$$

As $n$ is a square number, $\alpha_1, \alpha_2, \ldots, \alpha_k$ are even.

Then, the number of divisors of $n$ is

$$(\alpha_1 + 1)(\alpha_2 + 1) \cdots (\alpha_k + 1), \tag{23.10}$$

where $\alpha_1 + 1, \alpha_2 + 1, \ldots, \alpha_k + 1$ are odd, so their product (23.10) is also odd.

**Another Solution** Suppose that $n = m^2$, where $m$ is a natural number.

For every divisor $d$ of $n$,

$$n = d \times \frac{n}{d},$$

so $\frac{n}{d}$ is also a divisor of $n$. Then, we can pair the divisors $d$ and $\frac{n}{d}$ of $n$, and only $m$ is paired with itself ($\frac{n}{m} = m$). So, the number of divisors of $n$ is odd.

**Example 4.** How many natural numbers are there such that the remainder of 152 divided by them is 8?

**Analysis**   If the remainder of 152 divided by a natural number is 8, then this number must be a divisor of $152 - 8 = 144$. Then, the problem becomes: how many divisors of 144 are greater than 8?

**Solution**   $152 - 8 = 144 = 2^4 \times 3^2$.
   Totally, there are

$$(4+1) \times (2+1) = 15$$

divisors of 144. Among them, there are 6 divisors,

$$1, 2, 3, 4, 6, \text{ and } 8,$$

not greater than 8. So, there are $15 - 6 = 9$ numbers satisfying the problem requirement.

**Example 5.** Find the sum of all divisors of 72.

**Analysis**   If the decomposition of $n$ is as in (23.5), then in the expansion of (multiply out)

$$(1 + p_1 + p_1^2 + \cdots + p_1^{\alpha_1}) \times (1 + p_2 + p_2^2 + \cdots + p_2^{\alpha_2})$$
$$\times \cdots \times (1 + p_k + p_k^2 + \cdots + p_k^{\alpha_k}), \qquad (23.11)$$

every term is a number as in (23.6) and every two terms are different. Furthermore, any number in (23.6) will appear in the expansion. So, (23.11) represents the sum of all divisors of $n$.

**Solution**

$$72 = 2^3 \times 3^2,$$

so by formula (23.11), we have that the sum of all divisors of 72 is

$$(1 + 2 + 2^2 + 2^3) \times (1 + 3 + 3^2) = 15 \times 13 = 195.$$

   This is the same result as adding the 12 numbers in (23.2) together.

## Reading

### "Rich Numbers"

If a natural number $n > 1$ has more divisors than each of $1, 2, \ldots, n - 1$, then we call this $n$ a "rich number."
   Can you name the first 10 "rich numbers"?

## Exercises

1. Write down the numbers of divisors of 1–59, and fill them in the table of $d(n)$.

$n$

| tens digit<br>ones digit | 0 | 1 | 2 | 3 | 4 | 5 |
|---|---|---|---|---|---|---|
| 0 | | 10 | 20 | 30 | 40 | 50 |
| 1 | 1 | 11 | 21 | 31 | 41 | 51 |
| 2 | 2 | 12 | 22 | 32 | 42 | 52 |
| 3 | 3 | 13 | 23 | 33 | 43 | 53 |
| 4 | 4 | 14 | 24 | 34 | 44 | 54 |
| 5 | 5 | 15 | 25 | 35 | 45 | 55 |
| 6 | 6 | 16 | 26 | 36 | 46 | 56 |
| 7 | 7 | 17 | 27 | 37 | 47 | 57 |
| 8 | 8 | 18 | 28 | 38 | 48 | 58 |
| 9 | 9 | 19 | 29 | 39 | 49 | 59 |

$d(n)$

| tens digit<br>ones digit | 0 | 1 | 2 | 3 | 4 | 5 |
|---|---|---|---|---|---|---|
| 0 | | | | | | |
| 1 | | | | | | |
| 2 | | | | | | |
| 3 | | | | | | |
| 4 | | | | | | |
| 5 | | | | | | |
| 6 | | | | | | |
| 7 | | | | | | |
| 8 | | | | | | |
| 9 | | | | | | |

2. How many divisors does 105 have?

3. How many divisors does 4500 have?

4. What is the smallest natural number with exactly 12 different divisors?

5. The number of divisors of a natural number $n$ is represented by $d(n)$:

   (1) find $d(42)$;

   (2) find the smallest natural number $n$ such that $d(n) = 8$;

   (3) if $d(n) = 2$, what kind of number is $n$? What if $d(n) = 3$?

6. Prove that $d(n)$ must be even if the natural number $n$ is not a square number.

7. Suppose that $n \leqslant 60$, and for each positive integer $m$ smaller than $n$, we have $d(m) < d(n)$. Find the value(s) of $n$.

8. A natural number has 10 different divisors, and its prime factor is 3 or 5. How large can this natural number be?

9. From 1 to 100, what is the sum of all natural numbers with exactly three divisors?

10. Write down all natural numbers from 300 to 600 that have an odd number of divisors.

11. Find the sum of divisors of 240.

12. There are 96 apples in a basket. If you don't take out all of them at once and you don't take them out one by one, but you need to take out the same number of apples each time and take them all at the end. How many different ways are there?

# Chapter 24

# Positional Notation

The decimal positional notation we usually use has two characteristics:

(1) Use ten numbers (digits), namely

$$0, 1, 2, 3, 4, 5, 6, 7, 8, 9.$$

(2) Every ten carries one.

In addition to the decimal positional notation, there are other positional notations. For example, the Mayans in South America used the base-20 positional notation, and Europeans had the base-12 positional notation (1 "dozen" means 12) and the base-60 positional notation (1 h equals 60 min). The old scales in China are of the base-16 positional notation: 1 catty is equal to 16 taels, so the idiom "half catty eight taels" means that two people are of equal strength, and if the new scale is used, it should be "half catty five taels."

Binary is the most commonly used positional notation other than decimal positional notation.

Binary uses the two numbers 0 and 1, and "every two carries one."

In decimal, the 6 in 365 means $6 \times 10$, and the 3 means $3 \times 10^2$. Similarly, in binary, the first digit 1 in 111 means $1 \times 2^2$, the second 1 means $1 \times 2$, and the third 1 (that is, the first 1 from the right) means 1.

In order to avoid confusion, a number in the base-$g$ positional notation is often equipped with a parenthesis and a $g$ at the bottom right. For example, 111 in binary is recorded as $(111)_2$. Decimal numbers are represented as usual.

According to the above description,

$$(111)_2 = 1 \times 2^2 + 1 \times 2 + 1 = 7.$$

This is how we convert binary to decimal.

Similarly, we can consider the base-$g$ positional notation. In the base-$g$ positional notation, there are $g$ digits, which represent $0, 1, \ldots, g-1$ (when $g > 10$, you need to define several new digits yourself as the 10 numbers $0, 1, \ldots, 9$ are not enough), and "every $g$ carries one."

Generally, a number in base-$g$ can be written as $(\overline{a_0 a_1 \cdots a_n})_g$, whose decimal is

$$(\overline{a_0 a_1 \cdots a_n})_g = a_0 \times g^n + a_1 \times g^{n-1} + \cdots + a_{n-1} \times g + a_n. \qquad (24.1)$$

Equation (24.1) indicates that we use the sum of products to convert base-$g$ into decimal.

**Example 1.** Convert $(5326)_8$ into decimal.

**Solution**

$$\begin{aligned}
(5326)_8 &= 5 \times 8^3 + 3 \times 8^2 + 2 \times 8 + 6 \\
&= 2774.
\end{aligned}$$

Conversely, to convert decimal to base-$g$, we use division.

**Example 2.** Convert 2774 into octal (base-8).

**Solution**　Use division (this notation is often called short division) to get the following:

$$
\begin{array}{r|llll}
8 & 2 & 7 & 7 & 4 & & \text{Remainder} \\ \cline{2-5}
& 8 & 3 & 4 & 6 & \cdots\cdots & 6 \\ \cline{3-5}
& & 8 & 4 & 3 & \cdots\cdots & 2 \\ \cline{4-5}
& & & 5 & & \cdots\cdots & 3
\end{array}
$$

It can be seen that $2774 = 5 \times 8^3 + 3 \times 8^2 + 2 \times 8 + 6 = (5326)_8$.

We write the remainders from bottom to top (5 can be regarded as the remainder obtained from another division by 8), and obtain the number 5326 in octal.

Because binary has only two numbers, it can be simulated by two states (such as the switch in a circuit). Binary operations are also very simple, so they are widely used, especially in computers, which basically use binary and octal positional notations.

The arithmetic operations of binary are basically the same as those of decimal, except that in addition and subtraction, "every ten carries one" is changed to "every two carries one." In multiplication and division, there is only one statement of the multiplication formula: "one by one gets one." Of course, any number multiplied by 0 will be 0. There is no need to memorize the multiplication table.

**Example 3.** Compute

(1) $(111)_2 \times (101)_2$;
(2) $(100011)_2 \div (101)_2$.

**Solution**

(1) Use vertical multiplication (to make it simple, we omit $(\ )_2$):

$$
\begin{array}{ccccccc}
 & & & & 1 & 1 & 1 \\
 & & & \times & 1 & 0 & 1 \\
\hline
 & & & & 1 & 1 & 1 \\
 & & 1 & 1 & 1 & & \\
\hline
 & 1 & 0 & 0 & 0 & 1 & 1 \\
\end{array}
$$

So, $(111)_2 \times (101)_2 = (100011)_2$.

(2)

$$
\begin{array}{r}
1\ 1\ 1 \phantom{0000} \\
1\ 0\ 1\ )\ \overline{1\ 0\ 0\ 0\ 1\ 1} \\
1\ 0\ 1 \phantom{0000000} \\
\hline
1\ 1\ 1 \phantom{00000} \\
1\ 0\ 1 \phantom{00000} \\
\hline
1\ 0\ 1 \\
1\ 0\ 1 \\
\hline
0 \\
\end{array}
$$

So, $(100011)_2 \div (101)_2 = (111)_2$.

Not only decimal and octal but also any two positional notations can be converted to each other.

**Example 4.** Convert $(1532)_8$ into quaternary (base-4).

**Analysis** We can first convert octal into decimal and then convert the decimal into quaternary.

**Solution One**

$$(1532)_8 = 8^3 + 5 \times 8^2 + 3 \times 8 + 2$$
$$= 512 + 320 + 24 + 2$$
$$= 858.$$

```
4 | 8  5  8
4 | 2  1  4  ······  2
    4 | 5  3  ······  2
    4 | 1  3  ······  1
        3  ······  1
```

So, $(1532)_8 = (31122)_4$.

**Solution Two**    We can convert powers of 8 directly into powers of 4:

$$(1532)_8 = 8^3 + 5 \times 8^2 + 3 \times 8 + 2$$
$$= 2 \times 4^4 + (1+4) \times 4^3 + (2+4) \times 4 + 2$$
$$= 2 \times 4^4 + (4^4 + 4^3) + (4^2 + 2 \times 4) + 2$$
$$= 3 \times 4^4 + 4^3 + 4^2 + 2 \times 4 + 2$$
$$= (31122)_4.$$

**Solution Three**    As 8 and 4 are both powers of 2, we can use binary:

$$(1532)_8 = (1000000000)_2 + (101)_2 \times (1000000)_2 + (11)_2 \times (1000)_2 + (10)_2$$
$$= (1101011010)_2 = (31122)_4.$$

The last step of converting binary into quaternary is to first take every two digits of binary from right to left as a section:

$$11, 01, 01, 10, 10,$$

and each section becomes a number in quaternary:

$$10 \to 2, 01 \to 1, 11 \to 1 \times 2 + 1 = 3.$$

There are $2^n$ binary numbers of $n$ digits, the smallest is 0 and the largest is

$$2^{n-1} + 2^{n-2} + \cdots + 2 + 1 = 2^n - 1.$$

So, the binary numbers with no more than $n$ digits are exactly the $2^n$ integers from 0 to $2^n - 1$ in decimal.

According to the above analysis, if $n$ weights are used to weigh an object and the weights are only allowed to be placed on one side of the balance, then according to the above, we can use 1 gram, 2 grams, $2^2$ grams, $\ldots$, $2^{n-1}$ grams weights to weigh all integer weights no more than $2^n - 1$.

**Example 5.** Using a balance, what is the minimum number of weights needed to weigh all integers between 1 gram and 100 grams (the weights can be placed on either side of the balance)?

**Analysis** We first explain that four weights are not enough. Then, give an example that five weights can weigh from 1 gram to 100 grams.

**Solution** Using four weights whose weights are $a$, $b$, $c$, and $d$ grams, the weight that can be weighed is

$$|ax + by + cz + du| \qquad (24.2)$$

grams, where $x$ can take three values: 0 (that is, not using weight $a$), 1 (weight $a$ is placed on the right side), and $-1$ (weight $a$ is placed on the left side). The same is true for $y$, $z$, and $u$. So, (24.2) has at most

$$3 \times 3 \times 3 \times 3$$

different values. As one of them is 0 ($x = y = z = u = 0$), and when $x$, $y$, $z$, and $u$ change signs at the same time, the value of (24.2) remains unchanged; in fact, (24.2) can only give at most

$$\frac{1}{2}(3 \times 3 \times 3 \times 3 + 1) = 41$$

different values, that is, at most 41 kinds of weights can be weighed. To weigh 100 kinds of weights from 1 gram to 100 grams, at least five weights are required.

On the other hand, it is not difficult to verify that the weights of 1, 3, 9, and 27 can weigh 1–40 grams (for example, $37 = 27 + 9 + 3 - 2$). Adding a weight of 60 grams, we can weigh from 1 to 100 grams (for example, $56 = 60 - (3 + 1)$).

Of course, the set of weights 1, 3, 9, 27, and 60 is not the only set of solutions, and the weights from 1 gram to 100 grams can also be weighed with five weights of 1, 3, 9, 27, and 81 grams (weights up to $\frac{1}{2}(3^5 - 1) = 121$ grams can be weighed).

Remark: Weights can be placed on either side of the balance, which means instead of binary, ternary is used. But now, the ternary digits are not 0, 1, and 2, but 0, 1, and −1. We can regard the weights placed on the right side of the balance as positive and those placed on the left side as negative.

## Reading

### The Hundred Chickens Problem

There was a senior official in Qinghe County in the Southern and Northern Dynasties and the Northern Wei Dynasty (386–535 AD). A chicken seller lived opposite him. The senior official heard that the chicken seller had a very clever son. He wanted to test this kid, so he took out a hundred coins and wanted to buy a hundred chickens.

Given that each rooster is worth 5 coins, each hen is worth 3 coins, and three chicks are worth 1 coin. How many roosters, hens, and chicks are sold for exactly one hundred coins?

This problem was not easy. But it didn't bother the kid at the opposite house. He quickly sent 4 roosters, 18 hens, and 78 chicks for a total of 100 chickens and a total value of

$$4 \times 5 + 18 \times 3 + 78 \div 3 = 100$$

coins.

The next day, the senior official asked someone to send another one hundred coins, and he wanted to buy one hundred chickens, but the numbers must not be the same as yesterday. This is deliberately making things difficult. Unexpectedly, the kid immediately grabbed 8 roosters, 11 hens, and 81 chicks and sent them to the official's house. These 100 chickens were just worth

$$8 \times 5 + 11 \times 3 + 81 \div 3 = 100$$

coins.

The senior official was satisfied. A neighbor asked what if he wants to buy 100 chickens with 100 coins and the numbers are different from the previous two times?

The boy smiled and said that I would give him 12 roosters, 4 hens, and 84 chicks.

The boy was called Zhang Qiujian. Later, he published *The Mathematical Classic of Zhang Qiujian*.

## Exercises

1. Convert $(102102)_3$ into decimal.
2. Convert 308 into ternary.
3. Convert $(3714)_8$ into binary.
4. Convert $(111002220)_3$ into quinary (base-5).
5. Compute:
   (1) $(10010)_2 \div (110)_2 + (10101)_2 \div (11)_2$;
   (2) $(100100)_2 - (1011)_2 \times (11)_2 + (11011)_2$.
6. Compute:

$$[(10001000100)_2 - (100010001)_2] \div (1001)_2 \times (11)_2.$$

7. In which positional notation does $4 \times 41 = 314$?
8. The ternary representation of $x$ is 12112211122211112222. What is the first digit of the base-9 representation of $x$?
9. In base-12, use $t$ to represent 10 and use $e$ to represent 11. Find the square of $(\overline{eee})_{12}$.
10. Prove that when $g \geqslant 5$, $(1234321)_g$ is a square number.
11. A natural number is both $(\overline{abc})_9$ and $(\overline{cba})_7$. Find the value of this number in decimal.
12. A keeper of ordnance divides 1000 rounds of bullets into 10 boxes so that when he wants to take away any number of bullets between 1 and 1000, he can just take away a few boxes without opening them. How many bullets are in each box?

# Chapter 25

# Modular Arithmetic

The remainder of $13 \div 5$ is 3; the remainder of $28 \div 5$ is also 3. When the remainders of the two numbers 28 and 13 divided by 5 are the same, the difference between the two numbers, $28 - 13$, is divisible by 5.

Conversely, for two numbers, such as 28 and 13, if their difference $28 - 13$ is divisible by 5, then their remainders divided by 5 are the same.

Generally, given a positive integer $m$, if the difference $a - b$ of two integers $a$ and $b$ is divisible by $m$, then we say that $a$ and $b$ are congruent modulo $m$, denoted as

$$a \equiv b(\text{mod } m). \tag{25.1}$$

Equation (25.1) is simply $a = b + km$, where $k$ is an integer.

It is easy to know that congruence has the following properties:

(1) Reflexivity: $a \equiv a(\text{mod } m)$.
(2) Symmetry: If $a \equiv b(\text{mod } m)$, then $b \equiv a(\text{mod } m)$.
(3) Transitivity: If $a \equiv b(\text{mod } m)$ and $b \equiv c(\text{mod } m)$, then $a \equiv c(\text{mod } m)$.

Take the proof of transitivity as an example:

As $a \equiv b(\text{mod } m)$ and $b \equiv c(\text{mod } m)$, so $a = b + km$ and $b = c + hm$, where $k$ and $h$ are integers.

Therefore,

$$a = (c + hm) + km = c + (h + k)m,$$

that is, $a \equiv c(\text{mod } m)$.

It can also be proved that congruences have properties similar to equations:

(4) If $a \equiv b(\text{mod } m)$ and $c \equiv d(\text{mod } m)$, then $a \pm c \equiv b \pm d(\text{mod } m)$.

(5) If $a \equiv b(\bmod m)$ and $c \equiv d(\bmod m)$, then $ac \equiv bd(\bmod m)$.

(6) If $a \equiv b(\bmod m)$ and $n$ is a positive integer, then $a^n \equiv b^n(\bmod m)$.

But it should be noted that $a \equiv b(\bmod m)$ cannot be derived from $ka \equiv kb(\bmod m)$. For example, $54 \equiv 24(\bmod 15)$, but when both sides are divided by 6, the $9 \equiv 4(\bmod 15)$ is not true.

**Example 1.** If today is Sunday, what day will it be after 253 days?

**Solution**   Everyone knows that if today is Sunday, then every seven days, it will be Sunday again. As the remainder of $253 \div 7$ is 1,

$$253 \equiv 1(\bmod 7),$$

so after 253 days, it will be a Monday.

In this type of problem, the divisor (modulo) 7 is the "period." The quotient of $253 \div 7$ is not important; what matters is the remainder 1.

Example 2, as follows, is a bit more complicated, but if you think about it carefully, the rule is not difficult to identify.

**Example 2.** Stretch out your right hand, count from the thumb 1, 2, 3, 4, and 5 until the little finger, and then count back to the thumb and count forth, and so on (as shown in Figure 25.1). Which finger is it when counting to 2004?

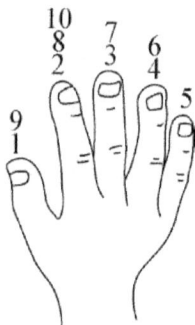

Fig. 25.1

**Solution**   Count from any finger and return to this finger every 8 counts, so 8 is the "period":

$$2004 \div 8 = 250 \cdots\cdots 4.$$

So, when counting to 2004, it is on the ring finger.

Remark: The key to this kind of "recurring" problem is to find the period of the recurrence and how many are left after an integer number of periods has passed.

**Example 3.** The New Year's Day in 2017 is a Sunday. What day is it $1993^{1999^{1997}}$ days after that?

**Solution**

$$1993 \equiv 5 \equiv -2(\text{mod } 7),$$

$$1993^3 \equiv (-2)^3 \equiv -8 \equiv -1(\text{mod } 7),$$

$$1993^6 \equiv (-1)^2 \equiv 1(\text{mod } 7).$$

Therefore, we should compute the remainder of $1999^{1997}$ divided by 6. As

$$1999 \equiv 1(\text{mod } 6),$$

we get

$$1999^{1997} \equiv 1(\text{mod } 6),\ 1993^{1999^{1997}} \equiv 1993^1 \equiv 5(\text{mod } 7).$$

Therefore, that day will be a Friday.

Remark: As $1993^6 \equiv 1(\text{mod } 7)$, so $1993^{6k} \equiv 1(\text{mod } 7)$, and we only need to find $1999^{1997} \equiv ?(\text{mod } 6)$. The value of $n$ which makes $a^n \equiv 1(\text{mod } m)$ is very useful. It can be proved that when $a$ is not a multiple of 7, it must be $a^6 \equiv 1(\text{mod } 7)$.

**Example 4.** Prove that $3^{1980} + 4^{1981}$ is divisible by 5.

**Solution** $3^2 \equiv 9 \equiv -1(\text{mod } 5)$, so

$$3^4 \equiv (-1)^2 \equiv 1(\text{mod } 5).$$

$$3^{1980} + 4^{1981} \equiv (3^4)^{495} + (-1)^{1981} \equiv 1 + (-1) \equiv 0(\text{mod } 5).$$

**Example 5.** Suppose that $x$ is an integer. Prove that $x^2 \equiv 0$ or $1(\text{mod } 4)$.

**Solution** If $x$ is even, then $x^2$ is divisible by 4, that is, $x^2 \equiv 0(\text{mod } 4)$. If $x$ is odd, $x = 2n + 1$, where $n$ is an integer, then

$$x^2 = (2n + 1)^2 = 4n^2 + 4n + 1 \equiv 1(\text{mod } 4).$$

Remark: The conclusion of this problem is very useful, so we should keep this in mind.

**Example 6.** The square of an integer is called a square number. Prove that there are no square numbers among

$$11, 111, 1111, \ldots.$$

**Solution**   As 100 is a multiple of 4, so

$$3 \equiv 11 \equiv 111 \equiv 1111 \equiv \cdots (\text{mod } 4).$$

From the previous example, a square number must be congruent to 0 or 1 modulo 4. We have that $11, 111, 1111, \ldots$ are not square numbers.

**Example 7.** Prove that the Diophantine equation

$$2x^2 - 5y^2 = 7 \tag{25.2}$$

has no integer solutions.

**Solution**   Suppose that (25.2) has integer solutions. As 7 is odd, so the left side of (25.2) is also odd, and $y$ is odd.

Taking mod 4 on both sides of (25.2), we find

$$2x^2 - y^2 \equiv -1 (\text{mod } 4). \tag{25.3}$$

By Example 5, we have $y^2 \equiv 1 (\text{mod } 4)$, so (25.3) is simply

$$2x^2 \equiv 0 (\text{mod } 4). \tag{25.4}$$

Therefore, $x$ must be even.

Take mod 8 on both sides of (25.2). As $2x^2 \equiv 0 (\text{mod } 8)$, so

$$3y^2 \equiv 7 (\text{mod } 8). \tag{25.5}$$

As $y$ is odd, $y = 2n + 1$. Then, we have

$$y^2 = (2n + 1)^2 = 4n^2 + 4n + 1 = 4n(n + 1) + 1.$$

As there is an even number between $n$ and $n + 1$, $4n(n + 1)$ is divisible by 8, and so

$$y^2 \equiv 1 (\text{mod } 8), \tag{25.6}$$

$$3y^2 \equiv 3 (\text{mod } 8). \tag{25.7}$$

Obviously, (25.5) and (25.7) contradict each other. Therefore, there are no integer solutions to (25.2).

**Remark:**   We first check the parity of the terms, that is, modulo 2, then modulo 4, and finally modulo 8, step by step. In order to find contradictions, we often choose different moduli. The most common moduli are prime numbers and their powers.

In Example 7, we find that the square of an odd number modulo 8 is 1, and this result goes one step further compared to the one we got in Example 5.

**Exercises**

1. Is there any natural number $n$ such that $n^2$ and $n$ are congruent to each other modulo 30? How many?

2. In decimal, $m = \overline{a_n a_{n-1} \cdots a_2 a_1}$, where $a_n$, $a_{n-1}, \ldots, a_1$ are digits. Prove that

$$m \equiv a_n + a_{n-1} + \cdots + a_2 + a_1 \pmod{9}.$$

3. Suppose that $A = 2001^{2002}$, $B$ is the digit sum of $A$, $C$ is the digit sum of $B$, and $D$ is the digit sum of $C$. Find the value of $D$.

4. Prove that $15 \mid 17^{2013} - 2$.

5. Prove that $11 \mid 10^{2013} + 23^{2015}$.

6. Suppose that $a$ is a positive integer. Prove that $a^5 \equiv a \pmod{10}$.

7. Prove that for any integer $a$, $10 \mid a^{2049} - a^{2013}$.

8. The remainder of $x$ divided by 3 is 2, and the remainder of $x$ divided by 4 is 1. Find the remainder of $x$ divided by 12.

9. The five-digit number $\overline{x679y}$ is divisible by 72. Find the digits $x$ and $y$.

10. Find the positive integer $n$ such that $(n+1) \mid (n^{2014} + 2006)$.

11. Find the last two digits of $2^{999}$.

12. Find the remainder of $1^5 + 2^5 + 3^5 + \cdots + 2013^5$ divided by 4.

# Chapter 26

# First-Degree Diophantine Equation with Two Unknowns

An equation with two or more unknown variables is called a Diophantine equation. In this chapter, we only discuss first-degree Diophantine equations with two unknown variables.

A first-degree Diophanine equation always has infinitely many solutions, but we only discuss the integer solutions or positive integer solutions of a Diophanine equation with integer coefficients. In this case, it may still have infinitely many solutions, or finitely many solutions, or even no solutions.

**Example 1.** Solve the equation $2x - 3y = 8$.

**Solution** From the original equation, it is easy to have $2x = 8 + 3y$ and $x = 4 + \frac{3}{2}y$.

Therefore, for any value of $y$, there is an $x$ ($= 4 + \frac{3}{2}y$) corresponding to it. The values of $x$ and $y$ satisfy the original equation, which form a solution to the original equation. That is, the solution of the original equation can be represented as

$$\begin{cases} x = 4 + \frac{3}{2}k, \\ y = k, \end{cases}$$

where $k$ is any number.

Remark: From the arbitrariness of the value of $y$, we know that the above Diophantine equation has infinitely many solutions. Generally, there are always infinitely many solutions for a first-degree Diophantine equation with two unknowns. The solution is similar to that in Example 1; that is, one of the unknowns is regarded as a parameter, and the other unknown is a function of it. The unknown regarded as a constant can be any number. For a first-degree Diophantine equation with two unknowns, we usually study its integer solutions.

In Example 1, it was only necessary to take $k$ as an even number, $k = 2h$, then $x$ and $y$ are both integers; that is, the integer solutions of Example 1 are

$$\begin{cases} x = 4 + 3h, \\ y = 2h, \end{cases}$$

where $h$ is any integer.

**Example 2.** Find the integer solutions of the equation $2x + 6y = 9$.

**Solution** As $2x + 6y = 2(x + 3y)$, we always have $2 \mid 2x + 6y$. But $2 \nmid 9$; therefore, no matter what integers $x$ and $y$ are, it is impossible that $2x + 6y$ is equal to 9. That is to say, the original equation has no integer solutions.

Remark: Example 2 tells us that not all first-degree equations with two unknowns have integer solutions. When does a first-degree equation with two unknowns have integer solutions, and when does it not? We have the following theorem.

**Theorem 1.** The necessary and sufficient condition for the integer coefficient equation $ax + by = c$ to have integer solutions is that the greatest common divisor $d$ of $a$ and $b$ divides $c$.

Theorem 1 tells us that if $d \mid c$, then the original equation has integer solutions. If $d \nmid c$, then the original equation does not have integer solutions.

**Example 3.** Find the integer solutions of the equation $4x + 10y = 34$.

**Solution** As the greatest common divisor of 4 and 10 is 2, and $2 \mid 34$, by Theorem 1, the original equation has integer solutions.

Dividing both sides by 2, we have $2x + 5y = 17$, and so

$$y = \frac{17 - 2x}{5} = 3 + \frac{2(1 - x)}{5}.$$

Therefore, to make $y$ an integer, $2(1 - x)$ must be a multiple of 5. As 2 and 5 are co-prime, $x - 1$ is a multiple of 5, that is, $x = 1 + 5k$, where $k$ is any integer. After substituting $x$ by $k$, we find

$$y = 3 - 2k.$$

That is, the integer solutions of the original equation are

$$\begin{cases} x = 1 + 5k, \\ y = 3 - 2k. \end{cases} \quad (k \text{ is any integer})$$

**Remark:** By Theorem 1, we have that if $ax + by = c$ has a solution, then the greatest common divisor $d$ of $a$ and $b$ divides $c$. Now, we can divide $d$ on both sides of the original equation and get $\frac{a}{d}x + \frac{b}{d}y = \frac{c}{d}$. Set $\frac{a}{d} = a_1$, $\frac{b}{d} = b_1$ and $\frac{c}{d} = c_1$. We get a first-degree equation with two unknowns $a_1 x + b_1 y = c_1$, which has the same solutions as the original equation. Then, the greatest common divisor of $a_1$ and $b_1$ is 1. Therefore, we only need to discuss the case of $d = 1$. We have the following theorem.

**Theorem 2.** If the greatest common divisor of $a$ and $b$ is 1 (that is, $a$ and $b$ are co-prime), $x_0$ and $y_0$ (we call it a special solution) give one solution to the integer coefficient first-degree Diophantine equation with two unknowns $ax + by = c$, and then all integer solutions (usually, we call them general solutions) of $ax + by = c$ are

$$\begin{cases} x = x_0 + bk, \\ y = y_0 - ak. \end{cases} \quad (k \text{ is any integer})$$

Therefore, when $d = 1$, $ax + by = c$ has solutions, and the key to solving this first-degree equation with two unknowns is to find a special solution $x_0$ and $y_0$.

**Example 4.** Find the integer solutions of the equation $2x + 3y = 5$.

**Solution** We can easily find that $x = 1$ and $y = 1$ is a solution to the equation. Also, as $(2, 3) = 1$, by Theorem 2, all integer solutions of the equation are

$$\begin{cases} x = 1 + 3k, \\ y = 1 - 2k. \end{cases} \quad (k \text{ is any integer}).$$

**Remark:** In this example, through observation, it is easy to find a solution. But sometimes, the special solution to the Diophantine equation is not easy to find, especially when the coefficients are large, such as the Diophantine equation $1999x + 105y = 1$.

In the following examples, we introduce common methods for finding special solutions of the Diophantine equations.

**Example 5.** Find the integer solutions of the equation $3x + 5y = 12$.

**Solution**   From $3x + 5y = 12$, we have

$$x = 4 - \frac{5}{3}y.$$

$x$ is an integer if and only if $3 \mid y$. Let $y = 3$, and then we have $x = 4 - \frac{5}{3} \times 3 = -1$; that is, $x = -1$ and $y = 3$ is a solution to the original equation. All integer solutions of the original equation are

$$\begin{cases} x = -1 + 5k, \\ y = 3 - 3k. \end{cases} \quad (k \text{ is any integer}).$$

**Example 6.** Find the integer solutions of the equation $3x + 5y = 31$.

**Solution**   From the original equation, we have

$$x = \frac{31 - 5y}{3},$$

that is,

$$x = 10 - 2y + \frac{1+y}{3}.$$

For $x$ to be an integer, $\frac{1+y}{3}$ must be an integer. Let $y = 2$, and we have $x = 10 - 2y + \frac{1+y}{3} = 10 - 4 + 1 = 7$. So, $x = 7$ and $y = 2$ is a solution to the original equation. Now, all solutions of the original equation are

$$\begin{cases} x = 7 + 5k, \\ y = 2 - 3k. \end{cases} \quad (k \text{ is any integer}).$$

Above, we have discussed the integer solutions of first-degree equations with two unknowns. Now, we study positive integer solutions of first-degree equations with two unknowns.

**Example 7.** Find the positive solutions of the equation $3x + 5y = 31$.

**Solution**   From Example 6, all integer solutions of $3x + 5y = 31$ are

$$\begin{cases} x = 7 + 5k, \\ y = 2 - 3k. \end{cases} \quad (k \text{ is any integer}).$$

To find the positive integer solutions to the original equation, we only need to make $x > 0$ and $y > 0$, which is the system of inequalities

$$\begin{cases} 7 + 5k > 0, \\ 2 - 3k > 0. \end{cases}$$

The solution to this system of inequalities is $-\frac{7}{5} < k < \frac{2}{3}$. Note that $k$ is an integer, so $k$ can only be $0$ or $-1$ in this range. Letting $k = 0$ and

$k = -1$, we find all positive integer solutions of the original equation as

$$\begin{cases} x = 7, \\ y = 2, \end{cases} \quad \begin{cases} x = 2, \\ y = 5. \end{cases}$$

Remark: To find the positive integer solutions of a first-degree Diophantine equation with two unknowns, you can first find its general solution. Then, set $x > 0$ and $y > 0$ to get a system of inequalities. Solve for the range of $k$ from the system of inequalities. In this range, find all integer solutions of $k$, and substitute them into the general solution. Then, we could obtain all positive integer solutions of this Diophantine equation.

**Example 8.** Find the positive integer solutions to the equation $5x - 3y = -7$.

**Solution** The original equation can be converted into $x = \frac{3y-7}{5}$, that is,

$$x = -2 + \frac{3(y+1)}{5}.$$

When $y = 4$, $x = 1$. That is,

$$\begin{cases} x = 1, \\ y = 4, \end{cases}$$

is an integer solution to the original equation. Now, all integer solutions of the original equation are

$$\begin{cases} x = 1 + 3k, \\ y = 4 + 5k. \end{cases} \quad (k \text{ is any integer})$$

Then, we set $x > 0$ and $y > 0$; that is, we have the system of inequalities

$$\begin{cases} 1 + 3k > 0, \\ 4 + 5k > 0. \end{cases}$$

The solution is $k > -\frac{1}{3}$. So, the integer positive solutions to the original equation are

$$\begin{cases} x = 1 + 3k, \\ y = 4 + 5k. \end{cases} \quad (k \text{ is any non-negative integer})$$

**Example 9.** Find the positive integer solutions to the equation $11x + 5y = 12$.

**Solution**   If there are positive integer solutions to the equation, then $x \geqslant 1$ and $y \geqslant 1$. Then,

$$11x + 5y \geqslant 11 + 5 = 16.$$

The right-hand side of the equation is 12, so the equation cannot hold and has no positive integer solutions.

Remark:   Generally, in the equation $ax + by = c$, if $a > 0$, $b > 0$, and $a + b > c$, then the equation has no positive integer solutions.

**Example 10.** Suppose that $a$ and $b$ are positive integers and that $\frac{2}{3}$, $\frac{a}{4}$, and $\frac{b}{6}$ are all simplified proper fractions. If $b$ is added to the numerators of $\frac{2}{3}$, $\frac{a}{4}$, and $\frac{b}{6}$, then the sum of the three new fractions is 6. Find the product of these three simplified proper fractions.

**Solution**   As given in the problem, we have $\frac{2+b}{3} + \frac{a+b}{4} + \frac{b+b}{6} = 6$, which can be simplified as

$$3a + 11b = 64. \tag{26.1}$$

It suffices to find the positive solutions of $3a + 11b = 64$.

By $3a + 11b = 64$, we have $a = \frac{64-11b}{3}$ and $a = 21 - 4b + \frac{1+b}{3}$.

Let $b = 2$, and we get $a - 14$. The Diophantine equation has an integer solution $\begin{cases} a = 14, \\ b = 2. \end{cases}$ Therefore, its integer solutions are

$$\begin{cases} a = 14 + 11k, \\ b = 2 - 3k \end{cases} \quad (k \text{ is any integer}).$$

Letting $a > 0$ and $b > 0$, we find the system of inequalities as

$$\begin{cases} 14 + 11k > 0, \\ 2 - 3k > 0. \end{cases}$$

The solution is $-\frac{14}{11} < k < \frac{2}{3}$. Thus, $k = 0$ or $k = -1$, and (26.1) has two positive integer solutions:

$$\begin{cases} a = 14, \\ b = 2, \end{cases} \quad \text{and} \quad \begin{cases} a = 3, \\ b = 5. \end{cases}$$

Note that $\frac{a}{4}$ and $\frac{b}{6}$ are simplified proper fractions, so $a = 3$ and $b = 5$ is the only solution. Therefore, the product we want to find is $\frac{2}{3} \times \frac{3}{4} \times \frac{5}{6} = \frac{5}{12}$.

Remark:   In the above solution to Example 10, we consider the general situations. In fact, there is a simpler solution as follows: as $\frac{a}{4}$ is a simplified fraction, $a$ can only be 1 or 3. Substitute them into (26.1); only when $a = 3$, $b$ is an integer. So, $a = 3$ and $b = 5$. The product we want to find is $\frac{5}{12}$.

# Reading

The Solution of the Hundred Chickens Problem

How do we solve the hundred chickens problem in Chapter 24 Reading? Suppose we buy $x$ roosters, $y$ hens, and $z$ chicks. Then,

$$\begin{cases} x + y + z = 100, & (26.2) \\ 5x + 3y + \dfrac{z}{3} = 100. & (26.3) \end{cases}$$

On $3 \times (26.3)$, we have

$$15x + 9y + z = 300. \qquad (26.4)$$

On $(26.4) - (26.2)$ and after simplifying, we get

$$7x + 4y = 100. \qquad (26.5)$$

As $x = 4$ and $y = 18$ is a special solution to $(26.5)$, the general solutions of $(26.5)$ are

$$\begin{cases} x = 4 + 4t, & (26.6) \\ y = 18 - 7t. & (26.7) \end{cases}$$

Substituting them in $(26.2)$, we find

$$z = 78 + 3t. \qquad (26.8)$$

It can be seen that every time the number of roosters increases by 4, the number of hens decreases by 7 and the number of chicks increases by 3, then another solution is obtained. So, all solutions are

$$\begin{cases} x = 4, & 8, & 12, \\ y = 18, & 11, & 4, \\ z = 78, & 81, & 84. \end{cases}$$

These are the three solutions given by Zhang QiuJian. The number of hens can no longer be reduced, but if the number of roosters is reduced to 0, then $18 + 7 = 25$ hens and $78 - 3 = 75$ chicks can also be regarded as a solution, and there are no other non-negative solutions.

**Exercises**

1. Determine whether each of the following first-degree equations with two unknowns has integer solutions and explain why.

   (1) $2x + 6y = 5$;

   (2) $4x + 6y = 8$;

   (3) $3x + 5 = 6y + 11$;

   (4) $x = \frac{11 - 2y}{3}$.

2. Find the solutions to the following first-degree equations with two unknowns:

   (1) $2x + 6y = 7$;

   (2) $-3x - 3 = 4y + 6$.

3. Find the integer solutions to the following first-degree equations with two unknowns:

   (1) $5x + 10y = 20$;

   (2) $3x - 4y = 7$;

   (3) $4x + 7y = 8$;

   (4) $13x + 30y = 4$.

4. Find the positive integer solutions to the following first-degree equations with two unknowns:

   (1) $11x + 15y = 20$;

   (2) $2x + 5y = 21$;

   (3) $5x - 2y = -3$;

   (4) $5x + 8y = 32$.

5. Try to write 100 as the sum of two positive integers, one of which is a multiple of 11 and the other is a multiple of 17.

6. Bob works for company $A$. A few months later, he also works for company $B$. Company $A$ pays him 470 yuan a month, and company $B$ pays him 350 yuan a month. At the end of the year, Bob gets a total salary of 7620 yuan. How many months has he worked in company $A$ and company $B$?

# Chapter 27

# The Drawer Principle

The drawer principle, also known as the pigeonhole principle, is a basic principle of combinatorial mathematics. The German mathematician Dirichlet was the first to explicitly state this principle. Therefore, the drawer principle is also called the Dirichlet principle.

To put three apples into two drawers, there must be two or more apples in one drawer. This well-known common sense is the drawer principle. Although the principle is simple, it is a common method to solve many existing problems, some of which can be very complicated

Common forms of the drawer principle are as follows:

**Principle 1.** If $n + 1$ apples are put into $n$ drawers, then there must exist a drawer with two or more apples.

**Principle 2.** To put $m$ apples into $n$ $(n < m)$ drawers, there must exist a drawer with at least $k$ apples, where

$$k = \begin{cases} \frac{m}{n}, & \text{when } n \text{ divides } m, \\ \left[\frac{m}{n}\right] + 1, & \text{when } n \text{ does not divide } m. \end{cases}$$

Here, $\left[\frac{m}{n}\right]$ represents the largest integer not greater than $\frac{m}{n}$, that is, the integer part of $\frac{m}{n}$.

**Principle 3.** If an infinite number of apples are put into a finite number of drawers, there must be a drawer containing an infinite number of apples.

**Example 1.** There are 50 students in a class. Some books need to be distributed to the students. At least how many books are needed to ensure that at least one student gets two or more books?

**Solution** The number of books should be more than the number of students; that is, at least $50 + 1 = 51$ books are needed to meet the requirement.

Remark: In this example, we can regard 50 students as 50 "drawers" and regard books as "apples."

**Example 2.** Eleven students borrow books from the teacher. The teacher has four types of books: $A$, $B$, $C$, and $D$. Each student can borrow either one or two books of different types. Try to prove that there must be two students who borrowed the same type(s) and number of books.

**Proof** For borrowing one book, there are 4 situations: $A$, $B$, $C$, and $D$. For borrowing two books, there are 6 situations: $AB$, $AC$, $AD$, $BC$, $BD$, and $CD$. There are 10 situations in total. Regard these 10 situations as ten "drawers" and the 11 students as "apples." According to the drawer principle, there are at least two students whose book(s) have the same type(s) and number.

Remark: To correctly understand the problem and reasonably design the "drawer" is the key to solving this kind of problem and also to mastering and applying the drawer principle. We should, based on the conditions and conclusions of the problems, comprehensively apply the relevant mathematical knowledge to grasp the most fundamental quantitative relationship and design the necessary drawers to solve the problem.

The above two examples are about students borrowing books. Sometimes students are regarded as "drawers" and books are regarded as "apples"; sometimes books are regarded as "drawers" and students are regarded as "apples." There are also other variations of making the "drawers" and "apples." Some commonly used "drawers" are given in the following.

**Example 3.** Can you fill in one of the three numbers 4, 5, and 6 in each small square of a $5 \times 5$ square (as shown in Figure 27.1) such that the sums of the five numbers in each row, column, and two diagonals are different? Why?

**Solution** It cannot be done.

After filling in the three numbers 4, 5, and 6 in each small square, the minimum sum of five numbers in a row, column, or diagonal is $4 \times 5 = 20$ (the five numbers are all 4), whereas the maximum is $6 \times 5 = 30$ (all five numbers are 6). As the sum of the five numbers is an integer, there are 11 different values (that is, all integer values between 20 and 30). Regard these 11 different integers as 11 "drawers." Regard the 12 sums of the 5 rows, 5 columns, and 2 diagonals as "apples." According to the drawer principle, there must be two or more equal sums.

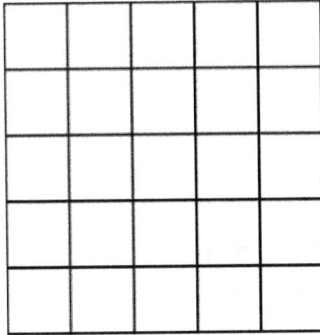

Fig. 27.1

Remark: In this example, we regard the values that the sums of five numbers can take as drawers.

**Example 4.** Nine points are randomly placed in a square with a side length of 2. If any three points are not on the same straight line, then there must be a triangle with three of these points as the vertices whose area does not exceed $\frac{1}{2}$.

**Proof** Divide the square with a side length of 2 into four identical small squares, as shown in Figure 27.2. Regard the four small squares as "drawers" and the 9 points as "apples." According to drawer principle 2, there is a small square which contains at least $[\frac{9}{4}] + 1 = 3$ of the 9 points (such as the three points $A$, $B$, and $C$ in the figure). As the three points $A$, $B$, and $C$ are in the square with a side length of 1, the area of the triangle $ABC$ does not exceed $\frac{1}{2}$. The proposition is proved.

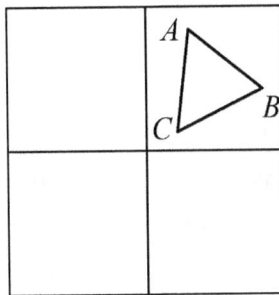

Fig. 27.2

**Remark:** In this example, we regard geometric figures (small squares) as drawers.

**Example 5.** There are six points $A$, $B$, $C$, $D$, $E$, and $F$ on a plane, and no three of them are collinear. Every two points are connected by a red line or a blue line. Prove that no matter how they are connected, there is at least one triangle whose three sides are of the same color.

**Proof** Choose any point from the six points, and suppose that it is point $A$. Among the five line segments connecting $A$ and the other five points, there must be at least $[\frac{5}{2}] + 1 = 3$ lines of the same color (we regard red and blue as drawers and the five line segments as "apples," in the drawer principle 2). Without loss of generality, we suppose that $AB$, $AC$, and $AD$ are red line segments. Now, if one of the three line segments $BC$, $BD$, or $CD$ is red (for example, $BC$ is red), then we have a triangle ($\triangle ABC$) whose three sides are red. Otherwise, $BC$, $BD$, and $CD$ are all blue, and $\triangle BCD$ is a triangle whose three sides are all blue.

**Remark:** This problem is the famous monochromatic triangle problem, and the conclusion is called Ramsey's Theorem. This problem can also be stated in another way: among any 6 people, there must be 3 people who know each other or 3 people who don't know each other.

**Example 6.** From $1, 2, \ldots, 2006$, at most how many numbers can be selected such that the difference between any two of the selected numbers is not equal to 6?

**Solution** Divide these 2006 numbers into 168 groups, each consisting of 12 consecutive numbers and the last group with the two numbers 2005 and 2006 only, as follows:

$$1, 2, 3, \ldots, 12;$$

$$13, 14, 15, \ldots, 24;$$

$$\vdots$$

$$1981, 1982, 1983, \ldots, 1992;$$

$$1993, 1994, 1995, \ldots, 2004;$$

$$2005, 2006.$$

Select the first 6 numbers in each group, and select the 2 numbers in the last group. Thus, we obtain a total of $6 \times 167 + 2 = 1004$ numbers.

Among these numbers, the difference between any two numbers in the same group is less than 6. Of two numbers in different groups, the difference is greater than 6. Therefore, the difference between any two numbers is not equal to 6.

On the other hand, if $1004 + 1 = 1005$ numbers are selected, then as $\frac{1005-2}{167} > \frac{1004-2}{167} = 6$, there must be one group in the first 167 groups such that, in this group, more than 6 numbers are selected. The 12 numbers in this group can be divided into 6 pairs, each pair having two numbers, $a$ and $a + 6$. When more than 6 numbers are selected, there must be two numbers from the same pair, and their difference is equal to 6. Therefore, a maximum of 1004 numbers can be selected.

**Another Solution**   Divide the numbers from 1 to 2006 into 6 "chains":

$$1 - 7 - 13 - 19 - 25 - \cdots - 1999 - 2005,$$

$$2 - 8 - 14 - 20 - 26 - \cdots - 2000 - 2006,$$

$$3 - 9 - 15 - 21 - 27 - \cdots - 2001,$$

$$4 - 10 - 16 - 22 - 28 - \cdots - 2002,$$

$$5 - 11 - 17 - 23 - 29 - \cdots - 2003,$$

$$6 - 12 - 18 - 24 - 30 - \cdots - 2004.$$

Each chain is an arithmetic sequence with a common difference of 6, that is, the difference between two adjacent terms is 6.

Select the 1st, 3rd, ... terms in each chain, and we select a total of $\frac{2004}{2} + 2 = 1004$ numbers (we select half of the terms in the third through sixth chains, as each chain has an even number of terms; we select one more term in the first and second chains, as each of them has an odd number of terms). The difference between any two of these numbers is not 6.

On the other hand, if more than 1004 are selected, then there must be a chain where two adjacent terms are selected with a difference of 6.

Remark:   The two solutions are essentially the same; think about it by yourself.

**Example 7.** Randomly select 51 numbers from the 100 natural numbers from 1 to 100. Prove that there must be two numbers among them whose difference is 50.

**Proof**   Divide the 100 natural numbers from 1 to 100 into 50 groups: $\{100, 50\}, \{99, 49\}, \{98, 48\}, \ldots, \{51, 1\}$. Each group has 2 numbers, and their difference is exactly 50. Regard these 50 groups as 50 drawers, and then at least 2 of the 51 random numbers are from the same drawer, and the difference between these two numbers is 50.

Remark: In the above example, we regard the pairs of numbers $a$ and $a + 50$ $(1 \leqslant a \leqslant 50)$ as drawers. If two numbers in the same drawer are selected, then the problem condition is met. This method of constructing drawers is quite common. Another example is as follows.

**Example 8.** Randomly select 6 numbers from the 10 natural numbers $1, 2, \ldots, 10$. Prove that there are two numbers among them such that one of them is a multiple of the other.

**Proof** Divide the 10 natural numbers $1, 2, \ldots, 10$ into five groups according to their odd factors:

$$A = \{1, 2, 4, 8\}, \quad B = \{3, 6\},$$
$$C = \{5, 10\}, \quad D = \{7\}, \quad E = \{9\}.$$

Regard these five groups as five "drawers." Select any 6 numbers from these five "drawers." According to the drawer principle, at least two numbers are from the same drawer, that is, in the same group. Since each of groups $D$ and $E$ contains only one number, these two numbers cannot appear in groups $D$ and $E$, and they must be in one of the three groups $A$, $B$, and $C$. In any of these three groups, the selected larger number must be a multiple of the smaller one.

**Example 9.** Select $n$ natural numbers randomly $(n \geqslant 1)$. Prove that among these $n$ natural numbers, either one number is a multiple of $n$, or the difference between some two numbers is a multiple of $n$.

**Proof** If one of the $n$ natural numbers is a multiple of $n$, then the conclusion is verified. Suppose that none of the $n$ natural numbers is a multiple of $n$. Then, when these natural numbers are divided by $n$, the remainders must be $1, 2, \ldots, n - 1$. Group the numbers with the same remainder into one category, which gives $(n - 1)$ categories in total. Regard these $(n - 1)$ categories as $(n - 1)$ "drawers" and the $n$ natural numbers as "apples." According to the drawer principle, there must be two numbers in the same category, that is, the remainders of these two numbers divided by $n$ are equal. Therefore, the difference between some two numbers is a multiple of $n$.

Remark: Taking the positive integer $n$ as the modulus (that is, the divisor), all integers can be divided into several categories according to the remainders; in other words, those integers with the same remainder are classified into one category. As the remainder can be $0, 1, \ldots, n-1$, all integers can be divided into $n$ categories, and these $n$ categories are called the residue class modulo $n$. Constructing drawers with the residue class modulo $n$ is one of the basic methods for solving problems.

The conclusion of Example 9 is also useful. Consider the following example.

**Example 10.** Prove that one of the 1999 numbers $1, 11, 111, \ldots, \underbrace{111 \cdots 11}_{1999}$

must be a multiple of 1999.

**Proof** From Example 9, there are two cases:

(1) There is a number that is a multiple of 1999.
    If so, the conclusion has been proved.
(2) The difference between two numbers is a multiple of 1999.
    Without loss of generality, suppose that these two numbers are $\underbrace{11 \cdots 11}_{i}$

and $\underbrace{11 \cdots 11}_{i+j}$ $(1 \leqslant i < i+j \leqslant 1999)$. Then,

$$\underbrace{11 \cdots 11}_{i+j} - \underbrace{11 \cdots 11}_{i} = \underbrace{11 \cdots 11}_{j}\underbrace{00 \cdots 00}_{i} = \underbrace{11 \cdots 11}_{j} \times 10^{i}$$

is a multiple of 1999. As 1999 and $10^i$ are co-prime, $\underbrace{11 \cdots 11}_{j}$ is a mul-

tiple of 1999.

So, the proposition is proved.

Remark: From the above example, we can see that the key to solving this kind of problem is "constructing drawers." The drawer principle is not difficult to understand, and you don't need advanced mathematical knowledge to solve the problems, but you do need more practice to use the drawer principle wisely.

## Reading

---

### Euclid and his *Elements*

The *Elements*, written by Euclid, is a compilation of mathematical knowledge of his time, consisting of a total of 13 volumes, covering plane and solid geometry, number theory, and incommensurable lines. It later became the most influential and widely circulated mathematical work and was translated into many other languages. In 1607, Xu Guangqi, a scholar in the Ming Dynasty of China, translated the first six volumes, together with Matteo Ricci, an Italian missionary, and translated the title of the book into *Elements of Geometry* (after another 250 years, Li Shanlan, a mathematician in the Qing Dynasty, and Alexander Wylie, an Englishman, completed the translation of the book).

*Elements* starts with the first few axioms (postulates), adopts the deductive method, and introduces new propositions in a logical order. It is a shining example of deductive reasoning. It constitutes the first mathematical axiom system in history. Therefore, Euclid is known as the father of geometry and has been synonymous with geometry for quite a long time.

Many people imitate the writing method of *Elements*. Examples include the *Mathematical Principles of Natural Philosophy* by scientist I. Newton (1642–1727), *Ethics* by philosopher B. Spinoza (1632–1677) (proved by the procedures of geometry), and the *Principles of Cartesian Philosophy*. Even Kang Youwei, the leader of the reformists in the late Qing Dynasty in China, prepared to use this method to deduce his doctrine.

Einstein believed that the development of modern Western science is based on two great achievements: one is the formal logic system represented by Euclidean geometry, which was invented by Greek philosophers; the other is the conclusion that it is possible to find causality through systematic experimentation, which was confirmed during the Renaissance (*Einstein Collected Works*, Volume 1, Commercial Press, 1957, p. 574).

---

### Exercises

1. In a class, randomly select 13 students. Prove that among the 13 students, at least two of them have the same zodiac sign.

2. A table tennis player hits the ball 65 times in 1 min. Try to prove that there is always a second in which he hits the ball more than once.

3. Randomly pick 5 points in an equilateral triangle with side length 1. Prove that there must be two points in it such that their distance is not more than $\frac{1}{2}$.

4. Arbitrarily given 11 natural numbers, try to prove that there are at least two of them such that their difference is a multiple of 10.

5. Randomly select 51 of the 100 natural numbers from 1 to 100. Prove that there are two numbers among the numbers selected such that one number is a multiple of the other.

6. Select 5 among the 8 numbers 1, 3, 5, ..., 15. Try to prove that among the 5 numbers selected, there are two numbers whose sum is 16.

7. Try to prove that among any 6 people, either 3 people know each other or 3 people don't know each other.

8. There are 3 red balls, 5 blue balls, and 7 white balls in a box. At least how many balls must be taken out to ensure that two of the balls taken out have the same color?

9. Given any 7 different integers, prove that there must be two integers whose sum or difference is a multiple of 10.

10. Put 11 potted flowers on a 20 m long concrete balcony. At least how many pairs of potted flowers have distances between each other not more than 2 m?

11. There are 3 red balls, 5 blue balls, and 7 white balls in a box. At least how many balls must be taken out to ensure that they include all three colors?

12. A total of 17 people communicate with each other and discuss three issues. Every communication between two people only discusses one of the issues. Prove that there are at least three people with whom they discuss the same issue in their communications.

13. $n$ natural numbers constitute a sequence as $a_1, a_2, \ldots, a_n$. Prove that there must be a number or a sum of several consecutive numbers in this sequence which is divisible by $n$.

14. Try to prove that among any 17 different positive integers, there must be several numbers among them such that they can be combined into a formula with only minus signs, multiplication signs, and brackets such that the result of the formula is a multiple of 21879.

15. Among any 52 natural numbers, there must be two numbers whose sum or difference is a multiple of 100.

16. Old classmates get together and shake hands with each other. Every two people shake hands at most once. Try to prove that there are at least two people such that they shake hands the same number of times.
17. A total of 2007 natural numbers are arbitrarily given. Prove that there must be several natural numbers among them such that their sum is a multiple of 2007 (a single number is also regarded as a sum).

# Solutions

# Solution 1

# Ingenious Computation of Rational Numbers

1. The original formula

$$= (31 - 22 + 4 + 11) + \left(\frac{2}{7} + \frac{5}{7}\right) + \left(-\frac{6}{13} + \frac{6}{13}\right)$$

$$= 24 + 1 = 25.$$

2. The original formula

$$= (5 - 3 - 7 - 3 + 8 - 3 - 2 + 6) + \left(\frac{6}{11} - \frac{4}{11} - \frac{2}{11}\right)$$

$$+ \left(-0.125 + \frac{1}{8}\right) + \left(-\frac{4}{7} - \frac{6}{7} + \frac{3}{7}\right) = 0.$$

3. The original formula

$$= \frac{7}{11} \times \frac{2}{5} \times \frac{3}{4} \times \frac{5}{7} \times \frac{11}{3} \times \frac{8}{13} = \frac{4}{13}.$$

4. The original formula

$$= 3.825 \times \left(\frac{1}{4} + 0.25 + \frac{1}{2}\right) - 1.825 = 2.$$

5. The original formula

$$= 3.6 \times \left(-0.25 + \frac{1}{2}\right) + 0.375 \times (1.1 - 3.5)$$

$$= 0.9 - 0.375 \times 2.4$$

$$= 0.9 - 3 \times 0.3 = 0.$$

6. The original formula
$$= \frac{1}{2 - \frac{1}{3 - \frac{5}{19}}} = \frac{1}{2 - \frac{19}{52}} = \frac{52}{85}.$$

7. The original formula
$$= (1 + 3 + 5 + 7 + 9) + \left( \frac{1}{2} + \frac{1}{4} + \frac{1}{8} + \frac{1}{16} + \frac{1}{32} \right)$$
$$= 5 \times 5 + \left( \frac{1}{2} + \frac{1}{4} + \cdots + \frac{1}{32} + \frac{1}{32} \right) - \frac{1}{32}$$
$$= 25 \frac{31}{32}.$$

8. The original formula
$$= \frac{1}{1999} \times (1 + 2 + \cdots + 1998)$$
$$= \frac{1}{1999} \times \frac{1998 \times (1998 + 1)}{2}$$
$$= 999.$$

9. The original formula
$$= (7 + 9 + \cdots + 99) + 101 - (7 + 9 + \cdots + 99) - 5$$
$$= 101 - 5 = 96.$$

10. The original formula
$$= 10 + 100 + 1000 + 10000 + 100000 + 1000000 - 6$$
$$= 1111110 - 6 = 1111104.$$

11. The original formula
$$= 3^{1998} \times (9 - 5 \times 3 + 6) = 0.$$

12. The original formula
$$= 1 - 1 + 1 - 1 = 0.$$

13. The original formula
$$= \frac{1}{4} \left( \frac{1}{5} - \frac{1}{9} + \frac{1}{9} - \frac{1}{13} + \cdots + \frac{1}{101} - \frac{1}{105} \right)$$
$$= \frac{1}{4} \times \left( \frac{1}{5} - \frac{1}{105} \right)$$
$$= \frac{1}{21}.$$

14. The original formula

$$= \left(2002\frac{1}{2} - 2001\frac{1}{3}\right) + \left(2000\frac{1}{2} - 1999\frac{1}{3}\right) + \cdots + \left(2\frac{1}{2} - 1\frac{1}{3}\right)$$

$$= \frac{7}{6} \times 1001 = \frac{7007}{6}.$$

15. The original formula

$$= \frac{1 \times 2 \times 3 \times (1 + 2 \times 2 \times 2 + 4 \times 4 \times 4 + 7 \times 7 \times 7)}{1 \times 3 \times 5 \times (1 + 2 \times 2 \times 2 + 4 \times 4 \times 4 + 7 \times 7 \times 7)} = \frac{2}{5}.$$

16. The original formula

$$= (1 + 2 + \cdots + 7) + \left(\frac{1}{2} - \frac{1}{3} + \frac{1}{3} - \frac{1}{4} + \cdots + \frac{1}{7} - \frac{1}{8}\right)$$

$$= \frac{(1 + 7) \times 7}{2} + \left(\frac{1}{2} - \frac{1}{8}\right) = 28\frac{3}{8}.$$

17. The original formula

$$= \frac{2}{1 \times 2} + \frac{2}{2 \times 3} + \frac{2}{3 \times 4} + \frac{2}{4 \times 5} + \cdots + \frac{2}{100 \times 101}$$

$$= 2 \times \left(1 - \frac{1}{2} + \frac{1}{2} - \frac{1}{3} + \cdots + \frac{1}{100} - \frac{1}{101}\right)$$

$$= 2 \times \left(1 - \frac{1}{101}\right)$$

$$= \frac{200}{101}.$$

# Solution 2

# Absolute Value

1. (1) ✓
   (2) When $a = 0$, $|a| = 0$, and it is not positive. So, we should fill with ✗.
   (3) When $m = 0$, $|m| = m$. So, we should fill with ✗.
   (4) ✓.
   (5) When $a = -3$, $b = 1$, $a < b$, but $|a| > |b|$. So, we should fill with ✗.
   (6) When $a$ is negative, $a + |a| = 0$, and it is not positive. So, we should fill with ✗.

2. As $-1 < x < 1$, $x + 1 > 0$ and $1 - x > 0$,
$$|x + 1| - |x - 1| = x + 1 - (1 - x) = 2x.$$

3. Since $a < 0$, the original formula$= \frac{2a + 3a}{|-3a - a|} = \frac{5a}{-4a} = -\frac{5}{4}$.

4. All integers whose absolute value is less than 100 are $0, \pm 1, \pm 2, \ldots, \pm 99$, for a total of $2 \times 99 + 1 = 199$ numbers. Their sum $= 0 + 1 + (-1) + 2 + (-2) + \cdots + 99 + (-99) = 0$.

5. The zero point of $x - \frac{1}{5}$ is $\frac{1}{5}$, and the zero point of $x + \frac{1}{5}$ is $-\frac{1}{5}$. If $x \leqslant -\frac{1}{5}$, then the original formula $= \frac{1}{5} - x - (\frac{1}{5} + x) = -2x$. If $-\frac{1}{5} < x \leqslant \frac{1}{5}$, then the original formula $= \frac{1}{5} - x + (\frac{1}{5} + x) = \frac{2}{5}$. If $x > \frac{1}{5}$, then the original formula $= x - \frac{1}{5} + (\frac{1}{5} + x) = 2x$.

6. As $a = \pm 5\frac{2}{3}$ and $b = \pm 1\frac{1}{3}$, there are four possible situations: $a - b = 5\frac{2}{3} - 1\frac{1}{3} = 4\frac{1}{3}$, $5\frac{2}{3} - (-1\frac{1}{3}) = 7$, $-5\frac{2}{3} - 1\frac{1}{3} = -7$, or $-5\frac{2}{3} - (-1\frac{1}{3}) = -4\frac{1}{3}$.

7. Not necessarily correct; for example, if $a = 1$ and $b = -2$, then $a > b$, but $|a| < |b|$.

8. As shown in Figure 2.4, $a = -c$ and $b < -c$, so
$$\text{the original formula} = 0 - (b + c) + (a + b) = a - c = a + a = 2a.$$

9. If $a = 0$, then $|a - b| = |-b| = |b| = |a| + |b|$. If $b = 0$, similarly, $|a - b| = |a| + |b|$. If both $a$ and $b$ are positive, then $|a| + |b| = a + b$, $|a - b| = a - b$, or $b - a$, so $|a - b| < |a| + |b|$. If both $a$ and $b$ are negative, $|a| + |b| = -a - b$, $|a - b| = a - b$, or $b - a$, so $|a - b| < |a| + |b|$. If $a$ is positive and $b$ is negative, then $|a| + |b| = a - b$ and $|a - b| = a - b$, so $|a - b| = |a| + |b|$. If $a$ is negative and $b$ is positive, then $|a| + |b| = b - a$ and $|a - b| = b - a$, so $|a - b| = |a| + |b|$.

To conclude, if at least one of $a$ and $b$ is zero or if $a$ and $b$ have different (opposite) signs, then $|a - b| = |a| + |b|$. And these are all cases where $|a - b| = |a| + |b|$ holds. So, the relationship between $a$ and $b$ should be $ab \leqslant 0$.

10. As $|a + b| = 0$ and $|a - b| = 0$, $a = b = 0$.
$$|a^{2005} + b^{2005}| + |a^{2005} - b^{2005}| = 0.$$

11. The zero point of $2x - 3$ is $\frac{3}{2}$, the zero point of $3x - 5$ is $\frac{5}{3}$, and the zero point of $5x + 1$ is $-\frac{1}{5}$.

If $x \leqslant -\frac{1}{5}$, then we have

the original formula $= -2x + 3 - 3x + 5 + 5x + 1 = 9$.

If $-\frac{1}{5} < x \leqslant \frac{3}{2}$, then we have

the original formula $= -2x + 3 - 3x + 5 - 5x - 1 = -10x + 7$.

If $\frac{3}{2} < x \leqslant \frac{5}{3}$, then we have

the original formula $= 2x - 3 - 3x + 5 - 5x - 1 = -6x + 1$.

If $x > \frac{5}{3}$, then we have

the original formula $= 2x - 3 + 3x - 5 - 5x - 1 = -9$.

12. The zero point of $2x - 4$ is 2, and the zero point of $3x - 6$ is also 2. If $x \leqslant 2$, then $|2x - 4| - 6 = 4 - 2x - 6 = -2 - 2x$, and its zero point is $x = -1$. If $x > 2$, then $|2x - 4| - 6 = 2x - 10$, and its zero point is $x = 5$.

If $x \leqslant -1$, the original formula $= |4 - 2x - 6| + 6 - 3x = -2 - 2x + 6 - 3x = 4 - 5x$; if $-1 < x \leqslant 2$, then the original formula $= |4 - 2x - 6| + 6 - 3x = 2 + 2x + 6 - 3x = 8 - x$; if $2 < x \leqslant 5$, then the original formula $= |2x - 4 - 6| + 3x - 6 = 10 - 2x + 3x - 6 = 4 + x$; and if $x > 5$, then the original formula $= |2x - 4 - 6| + 3x - 6 = 2x - 10 + 3x - 6 = 5x - 16$.

13. If $a > 0$, then $a + |a| = a + a = 2a$;
if $a = 0$, then $a + |a| = 0$;
if $a < 0$, then $a + |a| = a - a = 0$.

# Solution 3

# First-Degree Equation with One Variable

1. (1) $x = -7$.

    (2) $2x = -15$, $x = -\frac{15}{2}$.

    (3) $2(4 - 3x) = 3(5x - 6)$, $21x = 26$, $x = \frac{26}{21}$.

    (4) $(\frac{5}{11} - \frac{3}{13})x = \frac{1}{23} - \frac{1}{7}$, $\frac{32}{11 \times 13}x = -\frac{16}{7 \times 23}$, $x = -\frac{143}{322}$.

    (5) $(2 - \frac{2}{3} - \frac{1}{3} + \frac{1}{2})x = -\frac{4}{3} - \frac{1}{6}$, $\frac{3}{2}x = -\frac{3}{2}$, $x = -1$.

    (6) $\frac{1}{16}x = 2 + 2 + 1 + \frac{1}{2} + \frac{1}{4}$, $x = 92$.

2. (1) $(4m - 2)x = 9$. If $m \neq \frac{1}{2}$, then $x = \frac{9}{4m-2}$. If $m = \frac{1}{2}$, then the equation becomes $0 = 9$, so it has no solutions.

    (2) $(a - 4)x = b + 8$. If $a \neq 4$, then $x = \frac{b+8}{a-4}$. If $a = 4$ and $b \neq -8$, then the original equation becomes $0 = b + 8$, which has no solutions. If $a = 4$ and $b = -8$, then the original equation becomes $0 = 0$, which has infinitely many solutions, and each number is a solution to the original equation.

    (3) $(3a - 2)x = 9a^2 - 4$. If $a \neq \frac{2}{3}$, then $x = \frac{9a^2-4}{3a-2}$, which is simply $x = 3a - 2$ (for $(3a+2)(3a-2) = 9a^2 - 4$). If $a = \frac{2}{3}$, then the original equation becomes $0 = 0$, which has infinitely many solutions, and any number is a solution to the equation.

    (4) $3m(x + n) = 2(x + 2)$, $(3m - 2)x = 4 - 3mn$. If $m \neq \frac{2}{3}$, then $x = \frac{4-3mn}{3m-2}$. If $m = \frac{2}{3}$, then the original equation becomes $0 = 4 - 2n$. If $n \neq 2$, then the original equation has no solutions. If $n = 2$, then the original equation has infinitely many solutions, and any number is a solution to it.

247

3. The original equation can be converted into $(3a - 2b + 1)x = 5 - 6a$. To have infinitely many solutions, it must be $3a - 2b + 1 = 0$ and $5 - 6a = 0$. Using the second formula, we have $a = \frac{5}{6}$. Substitute it into the first formula, and we have $2b = \frac{5}{2} + 1$, $b = \frac{7}{4}$. If $a = \frac{5}{6}$ and $b = \frac{7}{4}$, then the original equation becomes $\frac{5}{6}(x+2) = \frac{5}{2}x + 5$, which has infinitely many solutions, and any number is its solution.

4. The original equation can be converted into $(2a - 3)x = -3 - 2a$. To make it have no solutions, it must be $2a - 3 = 0$, which is $a = \frac{3}{2}$. If $a = \frac{3}{2}$, then the original equation becomes $0 = -6$, which has no solutions.

5. (1) $(m^2 + m)x = m^2 - 1$. If $m \neq 0, -1$, then $x = \frac{m^2 - 1}{m^2 + m}$, which is $x = \frac{m-1}{m}$ (for $m^2 - 1 = (m+1)(m-1)$, $m^2 + m = m(m+1)$). If $m = 0$, then the original equation will be converted into $0 = -1$, which has no solutions. If $m = -1$, then the original equation will be converted into $0 = 0$, which has infinitely many solutions, and any number is its solution.

   (2) If $m + n \neq 0$ and $m \neq 0$, then $mx - n = 0$, $x = \frac{n}{m}$. If $m+n \neq 0$ and $m = 0$, then $n \neq 0$, and the equation will become $-n = 0$, which has no solutions. If $m + n = 0$, the original equation will become $0 = 0$, which has infinitely many solutions, and any number is its solution.

6. The original equation gives $(a - 2)x = -b - 3$. As it has two different solutions, $a - 2 = 0$ and $-b - 3 = 0$, then $a = 2$ and $b = -3$. So, $(a + b)^{2007} = (2 - 3)^{2007} = -1$.

7. As the equation is a first-degree equation, $a + 1 = 0$, which is simply $a = -1$, and then the original equation is $3x + 15 = 0$, whose solution is $x = -5$.

8. Suppose $\overline{a_1 a_2 \cdots a_n} = x$, then $12 \times \overline{2a_1 a_2 \cdots a_n 1} = 12(2 \times 10^{n+1} + 10x + 1)$, $21 \times \overline{1a_1 a_2 \cdots a_n 2} = 21(10^{n+1} + 10x + 2)$. So, we have $12(2 \times 10^{n+1} + 10x + 1) = 21(10^{n+1} + 10x + 2)$. After simplification, we have $(210 - 120)x = (24 - 21) \times 10^{n+1} - 21 \times 2 + 12$, which is simply $90x = 3 \times 10^{n+1} - 30$, and $x = \frac{10^n - 1}{3}$. As $(10^n - 1) \div 3 = \underbrace{99 \cdots 9}_{n} \div 3 = \underbrace{33 \cdots 3}_{n}$, we have $x = \underbrace{33 \cdots 3}_{n}$, which is $\overline{a_1 a_2 \cdots a_n} = \underbrace{33 \cdots 3}_{n}$.

   **Another Solution** Suppose $\overline{1a_1 a_2 \cdots a_n 1} = y$, then $12 \times (10^{n+1} + y) = 21(y + 1)$, $9y = 12 \times 10^{n+1} - 21$, $y = \frac{4 \times 10^{n+1} - 7}{3}$. Now,

$$(4 \times 10^{n+1} - 7) \div 3 = 3\underbrace{99 \cdots 9}_{n}3 \div 3 = 1\underbrace{33 \cdots 3}_{n}1,$$

and so $\overline{a_1 a_2 \cdots a_n} = \underbrace{33 \cdots 3}_{n}$.

# Solution 4

# System of First-Degree Equations

1.

$$\begin{cases} 3x + 5y = 15, & (4.1) \\ 2x - 3y = -4. & (4.2) \end{cases}$$

On $(4.1) \times 3$, we have

$$9x + 15y = 45. \qquad (4.3)$$

On $(4.2) \times 5$, we have

$$10x - 15y = -20. \qquad (4.4)$$

On $(4.3) + (4.4)$, we have $19x = 25$, $x = \frac{25}{19}$. Substitute it in $(4.1)$ to get $5y = 15 - \frac{75}{19}$, $y = \frac{42}{19}$.

So, the solution to the original system of equations is $\begin{cases} x = \frac{25}{19}, \\ y = \frac{42}{19}. \end{cases}$

2.

$$\begin{cases} \dfrac{x}{3} + \dfrac{y}{4} = 1, & (4.5) \\ 2x - 3y = 5. & (4.6) \end{cases}$$

On $12 \times (4.5)$, we find

$$4x + 3y = 12. \qquad (4.7)$$

On $(4.6) + (4.7)$, we get $6x = 17$, $x = \frac{17}{6}$. Substitute it in $(4.6)$ to find $3y = \frac{17}{3} - 5$, $y = \frac{2}{9}$.

So, the solution to the original system of equations is $\begin{cases} x = \frac{17}{6}, \\ y = \frac{2}{9}. \end{cases}$

3.
$$\begin{cases} x - 2y - 3 = 0, & (4.8) \\ 5x + 4y + 8 = 0. & (4.9) \end{cases}$$

By (4.8), we have

$$x = 2y + 3. \qquad (4.10)$$

Substitute it in (4.9), and after simplification, we find $14y = -23$, $y = -\frac{23}{14}$. Substitute it in (4.10), and we get $x = -\frac{2}{7}$.

So, the solution to the original system of equations is $\begin{cases} x = -\frac{2}{7}, \\ y = -\frac{23}{14}. \end{cases}$

4.
$$\begin{cases} x - 2y + 3z = 0, & (4.11) \\ 3x + 2y + 5z = 12, & (4.12) \\ 2x - 4y - z = -7. & (4.13) \end{cases}$$

On (4.11) + (4.12) and after simplification, we get

$$x + 2z = 3. \qquad (4.14)$$

On (4.12) × 2 + (4.13), we find

$$8x + 9z = 17. \qquad (4.15)$$

On (4.14) × 8 − (4.15), we have

$$7z = 7, \quad z = 1. \qquad (4.16)$$

Substitute it in (4.14) to get

$$x = 1. \qquad (4.17)$$

Substitute (4.16) and (4.17) in (4.11), and we have $y = 2$.

So, the solution to the original system of equations is $\begin{cases} x = 1, \\ y = 2, \\ z = 1. \end{cases}$

5.
$$\begin{cases} 2x - 3y + 4z = 12, & (4.18) \\ x - y + 3z = 4, & (4.19) \\ 4x + y - 3z = -2. & (4.20) \end{cases}$$

On $(4.19) + (4.20)$, we have

$$5x = 2, \quad x = \frac{2}{5}. \tag{4.21}$$

Substitute it in (4.18), and after simplification, we get

$$-3y + 4z = \frac{56}{5}. \tag{4.22}$$

Substitute (4.21) in (4.19), and after simplification, we find

$$-y + 3z = \frac{18}{5}. \tag{4.23}$$

On $3 \times (4.23) - (4.22)$, we get $5z = -\frac{2}{5}$, $z = -\frac{2}{25}$. Substituting it in (4.23), we have $y = -\frac{96}{25}$.

So, the solution to the original system of equations is $\begin{cases} x = \frac{2}{5}, \\ y = -\frac{96}{25}, \\ z = -\frac{2}{25}. \end{cases}$

6.
$$\begin{cases} x + 2y + 3z = 15, & (4.24) \\ \dfrac{x + 4z}{3} = \dfrac{y - 3z}{4} = 3. & (4.25) \end{cases}$$

On (4.25), we get

$$x + 4z = 9, \tag{4.26}$$

$$y - 3z = 12, \tag{4.27}$$

which are simply

$$x = 9 - 4z, \tag{4.28}$$

$$y = 12 + 3z. \tag{4.29}$$

Substitute (4.28) and (4.29) in (4.24), and after simplification, we have

$$5z = -18, \quad z = -\frac{18}{5}. \tag{4.30}$$

Substitute (4.30) in (4.28) to get $x = \frac{117}{5}$. Substitute (4.30) in (4.28), and we find $y = \frac{6}{5}$.

So, the solution to the original system of equations is $\begin{cases} x = \frac{117}{5}, \\ y = \frac{6}{5}, \\ z = -\frac{18}{5}. \end{cases}$

7.
$$\begin{cases} 2002x + 2003y = 6007, & (4.31) \\ 2003x + 2002y = 6008. & (4.32) \end{cases}$$

On (4.31) + (4.32), we have

$$4005x + 4005y = 12015, \quad x + y = 3. \tag{4.33}$$

On (4.32) − (4.31), we get

$$x - y = 1. \tag{4.34}$$

On (4.33) + (4.34), and after simplification, we have $x = 2$. Substituting it in (4.34), we have $y = 1$.

So, the solution to the original system of equations is $\begin{cases} x = 2, \\ y = 1. \end{cases}$

8.
$$\begin{cases} 3(x + 2) + 4y - 10 = 0, & (4.35) \\ 5(x - 2) - 6(y + 2) = 11. & (4.36) \end{cases}$$

Equation (4.35) is actually equivalent to

$$3x + 4y = 4. \tag{4.37}$$

Equation (4.36) is actually equivalent to

$$5x - 6y = 33. \tag{4.38}$$

On (4.37) × 5 − (4.38) × 3, we have $38y = -79$, $y = -2\frac{3}{38}$. Substituting it in (4.37), we have $3x = 4 + 8\frac{6}{19}$, $x = 4\frac{2}{19}$.

So, the solution to the original system of equations is $\begin{cases} x = 4\frac{2}{19}, \\ y = -2\frac{3}{38}. \end{cases}$

9.
$$\begin{cases} x + y + z = 2, & (4.39) \\ y + z + w = 3, & (4.40) \\ z + w + x = 5, & (4.41) \\ w + x + y = 8. & (4.42) \end{cases}$$

On $\frac{(4.39)+(4.40)+(4.41)+(4.42)}{3}$, we get

$$x + y + z + w = 6. \tag{4.43}$$

On (4.43) − (4.39), we have $w = 4$; on (4.43) − (4.40), we find $x = 3$; on (4.43) − (4.41), we get $y = 1$; on (4.43) − (4.42), we have $z = -2$.

So, the solution to the original system of equations is $\begin{cases} x = 3, \\ y = 1, \\ z = -2, \\ w = 4. \end{cases}$

10.

$$\begin{cases} 3x + 2y + 2z = 5, & (4.44) \\ 2x + 3y + 2z = 7, & (4.45) \\ 2x + 2y + 3z = 9. & (4.46) \end{cases}$$

On $(4.44) - (4.45)$, we get

$$x - y = -2. \tag{4.47}$$

On $(4.44) \times 3 - (4.46) \times 2$, we have

$$5x + 2y = -3. \tag{4.48}$$

On $(4.48) + 2 \times (4.47)$, we find

$$7x = -7, x = -1. \tag{4.49}$$

Substitute it in $(4.47)$ to get

$$y = 1. \tag{4.50}$$

Substitute $(4.49)$ and $(4.50)$ in $(4.44)$, and we have $z = 3$.

So, the solution to the original system of equations is $\begin{cases} x = -1, \\ y = 1, \\ z = 3. \end{cases}$

11.

$$\begin{cases} 3x - 2y = 3, & (4.51) \\ 7x - 4y = 7. & (4.52) \end{cases}$$

On $(4.52) - (4.51) \times 2$, we get $x = 1$. Substituting it in $(4.51)$, we have $y = 0$.

So, the solution to the original system of equations is $\begin{cases} x = 1, \\ y = 0. \end{cases}$

12.

$$\begin{cases} x + y - 3z = 2a, & (4.53) \\ x - 3y + z = 2b, & (4.54) \\ -3x + y + z = 2c. & (4.55) \end{cases}$$

On $\frac{(4.54)-(4.55)}{4}$, we find

$$x - y = \frac{b-c}{2}. \tag{4.56}$$

On $\frac{(4.54)\times 3+(4.53)}{4}$, we have

$$x - 2y = \frac{3b+a}{2}. \tag{4.57}$$

On $(4.56) - (4.57)$, we get

$$y = -b - \frac{c}{2} - \frac{a}{2}. \tag{4.58}$$

Substitute it in (4.56) to find

$$x = -c - \frac{a}{2} - \frac{b}{2}. \tag{4.59}$$

Substitute (4.58) and (4.59) in (4.55), and we have $z = -a - \frac{b}{2} - \frac{c}{2}$.
So, the solution to the original system of equations is
$$\begin{cases} x = -c - \frac{a}{2} - \frac{b}{2}, \\ y = -b - \frac{c}{2} - \frac{a}{2}, \\ z = -a - \frac{b}{2} - \frac{c}{2}. \end{cases}$$

13.
$$\begin{cases} x : y : z = 1 : 3 : 5, & (4.60) \\ x + y + z = 18. & (4.61) \end{cases}$$

By (4.60), we have

$$y = 3x, \tag{4.62}$$

$$z = 5x. \tag{4.63}$$

Substitute them in (4.61), and we have $x = 2$. Substitute it in (4.62) and (4.63), and we have $y = 6$ and $z = 10$.

So, the solution to the original system of equations is $\begin{cases} x = 2, \\ y = 6, \\ z = 10. \end{cases}$

14.
$$\begin{cases} \dfrac{x}{3} = \dfrac{y}{2} = \dfrac{z}{5}, & (4.64) \\ 2x + 3y - 4z = 8. & (4.65) \end{cases}$$

Suppose the value of (4.64) is $k$. Then, $x = 3k$, $y = 2k$, and $z = 5k$.
Substitute them in (4.65), and we have $-8k = 8$, $k = -1$.

So, the solution to the original system of equations is $\begin{cases} x = -3, \\ y = -2, \\ z = -5. \end{cases}$

15.

$$\begin{cases} \dfrac{4}{3x} + \dfrac{3}{2y} = 7, & (4.66) \\[3mm] \dfrac{5}{x} - \dfrac{6}{y} = 3. & (4.67) \end{cases}$$

Treat $\frac{1}{x}$ and $\frac{1}{y}$ as unknown variables. On $(4.66) \times 4 + (4.67)$, we find

$$\left(\frac{16}{3} + 5\right)\frac{1}{x} = 31, \quad \frac{1}{x} = 3.$$

Substituting it in (4.67), we have $\frac{6}{y} = 12$, $\frac{1}{y} = 2$.

So, the solution to the original system of equations is $\begin{cases} x = \frac{1}{3}, \\ y = \frac{1}{2}. \end{cases}$

# Solution 5

# Application of a System of First-Degree Equations

1. According to the given condition, we have $\begin{cases} -b = 3, \\ 3a - b = 9. \end{cases}$ So, we have $b = -3$ and $a = 2$.

2. According to the given condition, we have

$$\begin{cases} a + 3 - b = 3, & (5.1) \\ a \times (-2)^2 + 3 \times (-2) - b = 4, & (5.2) \end{cases}$$

which is simply

$$\begin{cases} a - b = 0, & (5.3) \\ 4a - b = 10, & (5.4) \end{cases}$$

So, $a = b = \frac{10}{3}$. If $x = 3$, then the value of the algebraic formula $= \frac{10}{3} \times 3^2 + 3 \times 3 - \frac{10}{3} = 35\frac{2}{3}$.

3. According to the given condition, we have

$$\begin{cases} 3x + 2y - 4 = 0, & (5.5) \\ 3y - 2x + 5 = 0. & (5.6) \end{cases}$$

On $(5.5) \times 3 - (5.6) \times 2$, we get $13x = 22$, $x = \frac{22}{13}$. Substituting it in $(5.5)$, we have $y = -\frac{7}{13}$. So, $x = \frac{22}{13}$ and $y = -\frac{7}{13}$.

4. According to the given condition, we have

$$\begin{cases} x - 3y + 6 = 0, & (5.7) \\ 4x - 2y - 3 = 0. & (5.8) \end{cases}$$

On $(5.8) - (5.7) \times 4$, we get $10y = 27$, $y = \frac{27}{10}$. Substituting it in $(5.7)$, we have $x = \frac{21}{10}$. So, $x = \frac{21}{10}$ and $y = \frac{27}{10}$.

5. Suppose that these two natural numbers are $x$ and $y$. Then, $(x+y)(x-y) = 71$. As 71 is a prime number, $x+y > x-y$ and $x+y > 0$, so

$$\begin{cases} x+y = 71, & (5.9) \\ x-y = 1. & (5.10) \end{cases}$$

On $\frac{(5.9)+(5.10)}{2}$, we find $x = 36$. Substituting it in (5.9), we have $y = 35$.

So, the two numbers we want are 36 and 35.

6. As $2x + y > 0$, $x - 2y > 0$. As $2x + y > x - 2y$, $\begin{cases} 2x+y = 7, \\ x-2y = 1. \end{cases}$

On solving it, we have $x = 3$ and $y = 1$.

7. According to the given condition, we have

$$\begin{cases} 2m - 2 = 4 - m, & (5.11) \\ m - n = 2n - 1. & (5.12) \end{cases}$$

By (5.11), we get $m = 2$. Substituting it in (5.12), we have $n = 1$.
So, $m = 2$ and $n = 1$.

8. According to the given condition, we have

$$\begin{cases} 3a + b = 11, & (5.13) \\ 5a + 6 = 11. & (5.14) \end{cases}$$

By (5.14), we have $a = 1$. Substituting it in (5.13), we have $b = 8$.
So, $a = 1$ and $b = 8$.

9. According to the given condition, we have

$$\begin{cases} a+b+c = -2, & (5.15) \\ 9a + 3b + c = 8, & (5.16) \\ a - b + c = -4. & (5.17) \end{cases}$$

On (5.15) $-$ (5.17), we find $2b = 2$, $b = 1$. Substituting it in (5.15) and (5.16), we have

$$a + c = -3, \qquad (5.18)$$

$$9a + c = 5. \qquad (5.19)$$

On (5.19) $-$ (5.18), we get $8a = 8$, $a = 1$. Substituting it in (5.18), we have $c = -4$.

So, $a = 1$, $b = 1$ and $c = -4$.

10. According to the given condition, we have

$$\begin{cases} 3a + 2b - c = 0, & (5.20) \\ 2a + b = 0, & (5.21) \\ 2b + c = 0. & (5.22) \end{cases}$$

By (5.21), we get

$$b = -2a. \tag{5.23}$$

By (5.22) and (5.23), we find

$$c = -2b = 4a. \tag{5.24}$$

Substitute (5.23) and (5.24) in (5.20), and we have $-5a = 0$, $a = 0$. Thus, by (5.23) and (5.24), we have $b = 0$ and $c = 0$.

11. As $5 = 5 \times 1 = (-5) \times (-1)$, we have

$$\begin{cases} 2x - y = 5, & (5.25) \\ x - 2y = 1; & (5.26) \end{cases}$$

or

$$\begin{cases} 2x - y = 1, & (5.27) \\ x - 2y = 5; & (5.28) \end{cases}$$

or

$$\begin{cases} 2x - y = -5, & (5.29) \\ x - 2y = -1; & (5.30) \end{cases}$$

or

$$\begin{cases} 2x - y = -1, & (5.31) \\ x - 2y = -5. & (5.32) \end{cases}$$

On solving them one by one, we have $\begin{cases} x = 3, \\ y = 1; \end{cases} \begin{cases} x = -1, \\ y = -3; \end{cases} \begin{cases} x = -3, \\ y = -1; \end{cases}$ and $\begin{cases} x = 1, \\ y = 3. \end{cases}$

12. By $xy = x + y$, we have $xy - x - y + 1 = 1$, which is simply $(x - 1)(y - 1) = 1$. Then, we have

$$\begin{cases} x - 1 = 1, & (5.33) \\ y - 1 = 1; & (5.34) \end{cases}$$

or

$$\begin{cases} x - 1 = -1, & (5.35) \\ y - 1 = -1. & (5.36) \end{cases}$$

For the former system of equations, we have $\begin{cases} x = 2, \\ y = 2. \end{cases}$ For the latter system of equations, we have $\begin{cases} x = 0, \\ y = 0. \end{cases}$ But we only need positive integer solutions, so $\begin{cases} x = 2, \\ y = 2. \end{cases}$

13. $(m + n)(m - n) = 60$, $m + n > m - n$, and $m + n > 0$. As $m + n$ and $m - n$ have the same parity, both of them are even (otherwise, their product will be odd, which cannot be equal to the even number 60). So, we have

$$\begin{cases} m + n = 30, & (5.37) \\ m - n = 2; & (5.38) \end{cases}$$

or

$$\begin{cases} m + n = 2 \times 5, & (5.39) \\ m - n = 2 \times 3. & (5.40) \end{cases}$$

For the former system of equations, we have $\begin{cases} m = 16, \\ n = 14. \end{cases}$ For the latter system of equations, we have $\begin{cases} m = 8, \\ n = 2. \end{cases}$

So, there are two solutions $\begin{cases} m = 16, \\ n = 14; \end{cases}$ and $\begin{cases} m = 8, \\ n = 2. \end{cases}$

14. According to the given condition, we have

$$\begin{cases} 2a + b = 5, & (5.41) \\ 2c + d = -3, & (5.42) \\ 11a - 6 = 5, & (5.43) \\ 11c - d = -3. & (5.44) \end{cases}$$

By (5.43), we get

$$a = 1. \tag{5.45}$$

Substitute it in (5.41) to find

$$b = 3. \tag{5.46}$$

On (5.42) + (5.44), we have

$$13c = -6, \quad c = -\frac{6}{13}. \tag{5.47}$$

Substituting it in (5.42), we have $d = -\frac{27}{13}$. So $a = 1$, $b = 3$, $c = -\frac{6}{13}$, and $d = -\frac{27}{13}$.

# Solution 6

# Setting up (Systems of) Equations to Solve Word Problems

1. Suppose the tens digit of this two-digit number is $x$, and then its ones digit will be $x+5$. According to the condition, we have $10x+(x+5) = 3(x+x+5)$, which gives $5x = 10$, $x = 2$. So, this two-digit number is 27.

   **Another Solution** As the ones digit of this two-digit number is 5 greater than its tens digit, this two-digit number can be 16, 27, 38, or 49. It is not hard to check that only 27 is 3 times the sum of its digits. So, this two-digit number is 27.

2. Suppose that the distance between $A$ and $B$ is $x$ km. Then, $\frac{x}{10} - \frac{x}{15} = \frac{10}{60}$. Multiplying by 30 on both sides, we have $3x - 2x = 5$, $x = 5$.

   **Answer** The distance between $A$ and $B$ is 5 km.

   **Another Solution** Suppose that it takes Alice $t$ minutes to arrive at $B$. Then, it takes Bob $t+10$ minutes. Considering the distance between $A$ and $B$, we have the equation $15t \div 60 = 10(t+10) \div 60$, so $5t = 100$, $t = 20$, $15t \div 60 = 5$.

   **Answer** The distance between $A$ and $B$ is 5 km.

3. Suppose that $C$ needs $x$ days to complete the project. Then, $5(\frac{1}{15} + \frac{1}{20}) + \frac{x}{24} = 1$, $x = 10$.

   **Answer** $C$ needs 10 days to complete the project.

4. Suppose that originally there is $x$ kg of sugar water. As the weight of sugar in the solution does not change, we have $x \times 40\% = (x+10) \times 10\%$, then $4x = x + 10$, $x = \frac{10}{3}$.

   **Answer**   The original sugar water is $\frac{10}{3}$ kg.

5. Suppose that the team losses $x$ games and wins $(11 - x)$ games. Then, $x + 2(11 - x) = 17$, $x = 5$.

   **Answer**   This team losses 5 games and wins 6 games.

6. Suppose that the speed of this ship in still water is $x$ km/h, and the speed of running water is $y$ km/h. Then, $\begin{cases} 4(x+y) = 60, \\ 5(x-y) = 60, \end{cases}$ which is simply $\begin{cases} x+y = 15, \\ x-y = 12, \end{cases}$ so we have $\begin{cases} x = 13.5, \\ y = 1.5. \end{cases}$

   **Answer**   The speed of this ship in still water is 13.5 km/h, and the speed of running water is 1.5 km/h.

7. Suppose that the tens digit is $x$ and the ones digit is $y$. Then, $\begin{cases} 10x + y = 9y + 6, \\ 10x + y = 11x + 1, \end{cases}$ which gives

$$\begin{cases} 5x - 4y = 3, & (6.1) \\ y - x = 1. & (6.2) \end{cases}$$

   On $(6.2) \times 5 + (6.1)$, we have $y = 8$. Substituting it in $(6.2)$, we have $x = 7$.

   **Answer**   This two-digit number is 78.

8. Suppose one person can manufacture $x$ products per day by using a machine and manufacture $y$ products per day by handwork. Then, we have

$$\begin{cases} x + 3y = 60, & (6.3) \\ 2x + 2y = 80. & (6.4) \end{cases}$$

   By $(6.4)$, we get

$$x + y = 40. \tag{6.5}$$

   On $(6.3) - (6.5)$, we find $2y = 20$, $y = 10$. Substituting it in $(6.5)$, we have $x = 30$, so

$$3x + y = 3 \times 30 + 10 = 100.$$

   **Answer**   If 3 people use machines and 1 person relies on handwork, then 100 products can be manufactured per day.

9. Suppose that we require $x$ g of the solution with a concentration of 25% and $y$ g of the solution with a concentration of 20%. Then, we have

$$\begin{cases} x + y = 100, & (6.6) \\ \dfrac{25}{100}x + \dfrac{20}{100}y = 100 \times \dfrac{22}{100}. & (6.7) \end{cases}$$

By (6.7), we get

$$5x + 4y = 440, \tag{6.8}$$

On $(6.8) - 4 \times (6.6)$, we have $x = 40$. Substituting it in (6.6), we have $y = 60$.

**Answer**    We require 40 g of the solution with a concentration of 25% and 60 g of the solution with a concentration of 20%.

10. Suppose that $A$ is $x$ years old and $B$ is $y$ years old. Then, we have

$$\begin{cases} \dfrac{1}{2}x = y - (x - y), & (6.9) \\ x + (x - y) + x = 63. & (6.10) \end{cases}$$

It gives

$$\begin{cases} 3x = 4y, & (6.11) \\ 3x - y = 63. & (6.12) \end{cases}$$

Substituting (6.11) in (6.12), we have $y = 21$. Substituting it in (6.11), we have $x = 28$.

**Answer**    $A$ is 28 years old, and $B$ is 21 years old.

11. Suppose that the length of the train is $x$ m, and the speed of the train is $y$ m/min. Then, we have

$$\begin{cases} 1 \cdot y = 1000 + x, & (6.13) \\ \dfrac{40}{60}y = 1000 - x. & (6.14) \end{cases}$$

On $(6.13) + (6.14)$, we find $\frac{5}{3}y = 2000$, $y = 1200$. Substituting it in (6.13), we have $x = 200$.

**Answer**    The length of the train is 200 m, and the speed of the train is 1200 m/min.

12. Suppose that the original numbers of people in groups $A$ and $B$ are $x$ and $y$, respectively. Then, we have

$$\begin{cases} x + y = 28, & (6.15) \\ (x+2) : (y+6) = 2 : 1. & (6.16) \end{cases}$$

By (6.16), we get

$$x + 2 = 2(y + 6). \qquad (6.17)$$

On $(6.15) - (6.17)$, we find $3y = 18$, $y = 6$. Substituting it in (6.15), we have $x = 22$.

**Answer**  The original numbers of people in groups $A$ and $B$ are 22 and 6, respectively.

13. Suppose that the person passed $x$ km at the speed of 15 km/h. Then, he passed $(x - 20)$ km at the speed of 10 km/h. The whole journey is $(2x - 20)$ km. Based on the time relationship, we have the equation $\frac{x}{15} + \frac{x-20}{10} = \frac{2x-20}{12.5}$, which is simply $\frac{x}{15} + \frac{x-20}{10} = \frac{4x-40}{25}$. Multiplying by 150 on both sides, we have $10x + 15(x - 20) = 6(4x - 40)$, $x = 60$, $2x - 20 = 2 \times 60 - 20 = 100$.

**Answer**  The whole journey is 100 km.

14. Suppose the speed of the passenger train is $5k$ m/s and the speed of the freight train is $3k$ m/s. According to the condition, we have $(5k - 3k) \times 60 = 280 + 200$, and on solving it, we have $k = 4$. Then, $5k = 20$ and $3k = 12$. The crossing time will be $\frac{280+200}{5k+3k} = 15(s)$.

**Answer**  The speed of the passenger train is 20 m/s, and the speed of the freight train is 15 m/s. If these two trains are running toward each other on parallel tracks, it will take 15 s for them to cross each other.

**Another Solution**  We assign the variables as above. Then, the crossing time is $\frac{280+200}{5k+3k} = \frac{280+200}{5k-3k} \times \frac{5-3}{5+3} = 60 \times \frac{1}{4} = 15(s)$.

When using this method, it is not necessary to find the value of $k$: no matter how long these two trains are, the crossing time for the trains running toward each other is $\frac{1}{4}$ the crossing time of them running in the same direction.

15. Suppose after deleting the first digit of the original number, the five-digit number is $x$. Then, the original number is $1 \times 10^5 + x$, and the new number is $10x + 1$. According to the condition, we have $10x + 1 = 3(10^5 + x)$, and $x = \frac{1}{7}(3 \times 10^5 - 1) = 42857$.

**Answer**  The original six-digit number is 142857.

16. Suppose that the speed on the first day is $x$ km/day. Starting from the second day, on each day, he walks $y$ km more than the day before. Then, the speed on the last day will be $(x + 9y)$ km/day.
According to the condition, we have

$$15x = x + (x + y) + (x + 2y) + \cdots + (x + 9y),$$

which is simply $5x = \frac{(1+9) \times 9}{2} y$, $x = 9y$. If he walks every day at the speed of the last day of the first walking method, it will take

$$15x \div (x + 9y) = 15x \div 2x = 7.5 (\text{days}).$$

**Answer**   It takes 7.5 days.

# Solution 7

# (System of) First-Degree Inequalities

1. (1) After rearranging the terms, we have $-11 > 5x$, so $x < -\frac{11}{5}$.

   (2) After eliminating the denominators and simplifying the inequality, we have $-x > 19$, so $x < -19$.

   (3) $(3 + \frac{1}{3} + \frac{1}{2})x < 3 + \frac{3}{2}$, so $x < \frac{27}{23}$.

   (4) $2x > 2$ and $x \neq 2$, so $x > 1$ and $x \neq 2$.

   (5) Multiply by 6 on both sides, and we have $2(2x-1)-6 \geqslant 3x+2+3x$, $-10 \geqslant 2x$, $x \leqslant -5$.

   (6) Multiply by 8 on both sides, and we have $40 - 4x \geqslant 28 - (4x + 1 - 2x - 4)$. After simplification, we have $9 \geqslant 2x$, $x \leqslant \frac{9}{2}$.

2. (1)
$$\begin{cases} 2x - 2 > \dfrac{1}{2}(x - 7) + 1, & (7.1) \\[2mm] \dfrac{2x - 5}{3} + 3 < \dfrac{x}{5} + 2. & (7.2) \end{cases}$$

   By (7.1), we find $3x > -1$, $x > -\frac{1}{3}$. By (7.2), we get $7x < 10$, $x < \frac{10}{7}$.

   The solution of the system of inequalities is $-\frac{1}{3} < x < \frac{10}{7}$.

   (2)
$$\begin{cases} 4x + \dfrac{2}{3} < \dfrac{x - 8}{5} + 2, & (7.3) \\[2mm] 2 - \dfrac{x}{3} > 3 - \dfrac{x}{2}. & (7.4) \end{cases}$$

269

By (7.3), we get

$$19x < -\frac{4}{3}, \quad x < -\frac{4}{57}. \tag{7.5}$$

By (7.4), we find

$$\frac{x}{6} > 1, \quad x > 6. \tag{7.6}$$

As (7.5) and (7.6) are contradicting, the original system of inequalities has no solutions.

(3)
$$\begin{cases} 1 < \dfrac{2x-5}{3} < 3, & (7.7) \\ \left(x - \dfrac{2}{3}\right) + 5 \geqslant 2x - \dfrac{3}{2}. & (7.8) \end{cases}$$

By (7.7), we have $4 < x < 7$. By (7.8), we have $5\frac{5}{6} \geqslant x$. So, the solution of the system of inequalities is $4 < x \leqslant 5\frac{5}{6}$.

(4)
$$\begin{cases} x - 1 > -3, & (7.9) \\ \dfrac{x}{2} - 1 < \dfrac{x}{3}, & (7.10) \\ 3 < 2(x-1) < 10. & (7.11) \end{cases}$$

By (7.9), we have $x > -2$. By (7.10), we find $x < 6$. By (7.11), we get $\frac{5}{2} < x < 6$. So, the solution of the system of inequalities is $\frac{5}{2} < x < 6$.

(5)
$$\begin{cases} 2x + 3 < 9 - x, & (7.12) \\ 6x - 1 < 5, & (7.13) \\ 2 - x \leqslant 3x + 7. & (7.14) \end{cases}$$

By (7.12), we have $x < 2$. By (7.13), we have $x < 1$. By (7.14), we have $-\frac{5}{4} \leqslant x$.

So, the solution of the system of inequalities is $-\frac{5}{4} \leqslant x < 1$.

3. The solution of the equation is $x = 1 - \frac{19}{2}k$.

   (1) As the equation has a positive solution, $1 - \frac{19}{2}k > 0$, we get $k < \frac{2}{19}$.

   (2) As the equation has a negative solution, the direction of the inequality sign in (1) will change, which is $k > \frac{2}{19}$.

4. The original system of inequalities can be converted into

$$\begin{cases} (a-1)x > 2a - 3, & (7.15) \\[2mm] x > \dfrac{8}{9}. & (7.16) \end{cases}$$

If $a - 1 > 0$, which is $a > 1$, then by (7.15), we find $x > \frac{2a-3}{a-1}$. Now, solve the inequality $\frac{8}{9} > \frac{2a-3}{a-1}$, which is $8a - 8 > 18a - 27$, and we have $\frac{19}{10} > a$. So, when $\frac{19}{10} \geqslant a > 1$, $x > \frac{8}{9}$. When $a > \frac{19}{10}$, $x > \frac{2a-3}{a-1}$.

If $a = 1$, then (7.15) will become $0 > -1$, which of course holds. The solution is $x > \frac{8}{9}$.

If $a < 1$, then by (7.15), we get $x < \frac{2a-3}{a-1}$. Now, solve the inequality $\frac{8}{9} < \frac{2a-3}{a-1}$, which is (pay attention to $a - 1 < 0$) $8a - 8 > 18a - 27$, and we have $\frac{19}{10} > a$. So, when $a < 1$, $\frac{8}{9} < x < \frac{2a-3}{a-1}$.

To conclude, when $a > \frac{19}{10}$, $x > \frac{2a-3}{a-1}$; when $\frac{19}{10} \geqslant a \geqslant 1$, $x > \frac{8}{9}$; when $a < 1$, $\frac{8}{9} < x < \frac{2a-3}{a-1}$.

5. The original equation is simply $(3k - 5)x = 6 - 6k$. When $k = \frac{5}{3}$, the equation will become $0 = -4$ with no solutions.
   When $k \neq \frac{5}{3}$,

$$x = \frac{6 - 6k}{3k - 5}. \qquad (7.17)$$

Then, we try to solve $\frac{6-6k}{3k-5} \leqslant 3$.

When $k > \frac{5}{3}$, $6 - 6k \leqslant 3(3k - 5)$, $k \geqslant \frac{7}{5}$. So, we have $k > \frac{5}{3}$.

When $k < \frac{5}{3}$, $6 - 6k \geqslant 3(3k - 5)$, $k \leqslant \frac{7}{5}$. So, we have $k \leqslant \frac{7}{5}$.

To conclude, $k > \frac{5}{3}$ or $k \leqslant \frac{7}{5}$.

6. 
$$\frac{1}{2^2} + \frac{1}{3^2} + \cdots + \frac{1}{n^2} < \frac{1}{1 \times 2} + \frac{1}{2 \times 3} + \cdots + \frac{1}{(n-1)n}$$

$$= \left(1 - \frac{1}{2}\right) + \left(\frac{1}{2} - \frac{1}{3}\right) + \cdots + \left(\frac{1}{n-1} - \frac{1}{n}\right)$$

$$= 1 - \frac{1}{n} = \frac{n-1}{n}.$$

# Solution 8

# Multiplication and Division of Polynomials with Integer Coefficients

1. (1) The original formula $= 6x^3 + x^2 - 8x + 3$. We can do mental calculation or compute by polynomial long multiplication. In the following long multiplication, we arrange the polynomial in the descending powers, omit the $x$ symbols, and only write down the coefficients.

$$
\begin{array}{r}
3 + 2 - 3 \\
\times \quad 2 - 1 \\
\hline
- 3 - 2 + 3 \\
6 + 4 - 6 \quad\ \\
\hline
6 + 1 - 8 + 3
\end{array}
$$

(2) The original formula $= 20x^7 - 8x^6 - 26x^5 + 8x^4 + 4x^3 + 2x^2 + 2x - 2$.

$$
\begin{array}{r}
4 \ +0 \ -6 \ +0 \ +2 \\
\times \quad 5 \ -2 \ +1 \ -1 \\
\hline
-4 \quad\ +6 \quad -2 \\
4 \quad -6 \quad +2 \quad\ \\
-8 \quad +12 \quad -4 \quad\quad \\
20 \quad -30 \quad +10 \quad\quad\quad \\
\hline
20 \ -8 \ -26 \ +8 \ \ +4 \ \ +2 +2 -2
\end{array}
$$

Pay attention to the 0's and the blank spaces denoting the missing items.

(3) The original formula $= (a^2 + 2ab + b^2) - (a^2 - 2ab + b^2) = 4ab$.

(4) The original formula $= a^3 + 3a^2b + 3ab^2 + b^3 - 3ab(a+b) = a^3 + b^3$.

273

(5) The original formula $= a^3 + ab^2 + ac^2 - a^2b - abc - ca^2 + ba^2 + b^3 + bc^2 - ab^2 - b^2c - abc + ca^2 + cb^2 + c^3 - abc - bc^2 - ac^2 = a^3 + b^3 + c^3 - 3abc.$

(6)

$$
\begin{array}{r}
3 - 13 \\
1 + 3 - 1 \overline{)\,3 - 4 + 5 - 1} \\
\underline{3 + 9 - 3\phantom{00000}} \\
-13 + 8\ \ -1 \\
\underline{-13 - 39 + 13} \\
47 - 14
\end{array}
$$

The quotient is $3x - 13$, and the remainder is $47x - 14$. When there is only one variable, similar to multiplication, we only need to write down the coefficients (omit the letters).

(7)

$$
\begin{array}{r}
\dfrac{5}{2} - \dfrac{5}{4} - \dfrac{23}{8} \\
2 + 1 \overline{)\,5 + 0\ -7 + 1} \\
5 + \dfrac{5}{2}\phantom{000000} \\
\hline
-\dfrac{5}{2} - 7\phantom{000} \\
-\dfrac{5}{2} - \dfrac{5}{4}\phantom{00} \\
\hline
-\dfrac{23}{4} + 1 \\
-\dfrac{23}{4} - \dfrac{23}{8} \\
\hline
\dfrac{31}{8}
\end{array}
$$

The quotient is $\frac{5}{2}x^2 - \frac{5}{4}x - \frac{23}{8}$, and the remainder is $\frac{31}{8}$.

(8)

$$
\begin{array}{r}
1 - 1 + 1 \\
1 + 1 \overline{)\,1 + 0 + 0 + 1} \\
\underline{1 + 1\phantom{00000000}} \\
-1 + 0\phantom{0000} \\
\underline{-1 - 1\phantom{0000}} \\
1 + 1\phantom{00} \\
\underline{1 + 1\phantom{00}} \\
0
\end{array}
$$

The quotient is $x^2 - x + 1$, and the remainder is 0. This problem can also be solved using the multiplication formula $(x+1)(x^2 - x + 1) = x^3 + 1$, and the result follows directly.

(9) The original formula
$$= (a^2 - b^2)(a^3 + b^3) \div (a^2 + 2ab + b^2)$$
$$= (a+b)(a-b)(a+b)(a^2 - ab + b^2) \div (a+b)^2$$
$$= (a-b)(a^2 - ab + b^2)$$
$$= a^3 - a^2 b + ab^2 - a^2 b + ab^2 - b^3$$
$$= a^3 - 2a^2 b + 2ab^2 - b^3.$$

(10) The original formula
$$= (7x^2 + 3x) \div (7x + 3) \times (6x + 3) \div (2x + 1)$$
$$= x(7x + 3) \div (7x + 3) \times 3(2x + 1) \div (2x + 1)$$
$$= 3x.$$

2. (1) The original formula
$$= 2000^2 - (2000 - 1)(2000 + 1)$$
$$= 2000^2 - (2000^2 - 1)$$
$$= 1.$$

(2) The original formula
$$= (2-1)(2+1)(2^2 + 1) \cdots (2^{2^n} + 1)$$
$$= (2^2 - 1)(2^2 + 1) \cdots (2^{2^n} + 1)$$
$$= (2^4 - 1) \cdots (2^{2^n} + 1)$$
$$= \cdots = (2^{2^n} - 1)(2^{2^n} + 1)$$
$$= 2^{2^{n+1}} - 1.$$

3. $x^{128} + x^{110} - x^{32} + x^8 + x^2 - x = (x^{128} - 1) + (x^{110} - 1) - (x^{32} - 1) + (x^8 - 1) + (x^2 - 1) - (x - 1) + 2$. As $x^{128} - 1$, $x^{110} - 1$, $x^{32} - 1$, $x^8 - 1$, $x^2 - 1$, and $x - 1$ are all divisible by $x - 1$. So, the remainder is 2.

4. The original formula $= (x^{111} + 1) - (x^{31} + 1) + (x^{13} + 1) + (x^9 + 1) - (x^3 + 1) - 1$, and $x + 1 \mid x^n + 1$ (where $n$ is odd), so the remainder is $-1$.

5. Suppose that the two adjacent odd numbers are $2n-1$ and $2n+1$ (where $n$ is an integer). Then, the difference between their squares $(2n+1)^2 - (2n-1)^2 = 8n$ is a multiple of 8; their sum is $(2n+1) + (2n-1) = 4n$. So, the difference between their squares is twice their sum.

6. From Exercise 1(5), $(a + b + c)(a^2 + b^2 + c^2 - ab - bc - ca) = a^3 + b^3 + c^3 - 3abc$. So, when $a + b + c = 0$, $a^3 + b^3 + c^3 - 3abc = 0$, which is $a^3 + b^3 + c^3 = 3abc$.

7. $(x + y - 2z) + (y + z - 2x) + (z + x - 2y) = 0$, then according to the last problem (let $a = x + y - 2z$, $b = y + z - 2x$, and $c = z + x - 2y$), we have $(x + y - 2z)^3 + (y + z - 2x)^3 + (z + x - 2y)^3 = 3(x + y - 2z)(y + z - 2x)(z + x - 2y)$.

# Solution 9

# Line Segments

1. $AD + BE + CF > AP + BP + CP$. By Example 1, we have

$$AP + BP + CP > \frac{1}{2}(AB + BC + CA).$$

Thus, $AD + BE + CF > \frac{1}{2}(AB + BC + CA)$.

2. Extend $BP$ to intersect $AC$ at $D$.

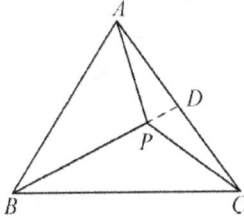

Fig. 9.1

In $\triangle ABD$, $AB + AD > BD = BP + PD$.

In $\triangle CPD$, $PD + DC > PC$.

Adding the above two formulas and eliminating $PD$, we have

$$AB + AD + DC > BP + PC,$$

that is, $AB + AC > PB + PC$.

For the same reason, we have

$$AC + BC > PA + PB,$$
$$AB + BC > PA + PC.$$

Adding the three formulas and then dividing the result by 2, we get the conclusion.

3. As shown in Figure 9.2, the grazing route is $A \to B \to C \to A$.

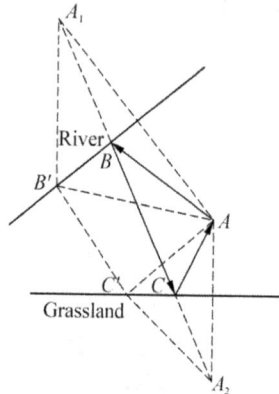

Fig. 9.2

Here, $A_1$ is the symmetric point of $A$ with respect to the river bank, and $A_2$ is the symmetric point of $A$ with respect to the grassland. Connect $A_1 A_2$, and suppose that it intersects the riverbank and the grassland at points $B$ and $C$, respectively. Now, the length of the grazing route $AB + BC + CA = A_1 B + BC + CA_2 = A_1 A_2$, which is simply the length of $A_1 A_2$. For an arbitrary point $B'$ on the river bank and an arbitrary point $C'$ on the grassland, by symmetry, $AB' = A_1 B'$ and $AC' = A_2 C'$. So, $AB' + B'C' + C'A = A_1 B' + B'C' + C'A_2 \geqslant A_1 A_2$, which is $AB' + B'C' + C'A \geqslant AB + BC + CA$. Thus, the given route is the shortest.

4. Draw a line through $A$ perpendicular to the river bank, and choose a point $E$ on this line which makes $AE$ equal to the width of the river. Join $BE$, and suppose that it intersects the river bank on the same side of $B$ at point $D$. Build the bridge $CD$ at $D$. Build the road connecting $A$ and $C$, and also build the road connecting $B$ and $D$. Then, the total length of the road is the shortest.

For any bridge $C_1D_1$, as $C_1D_1$ is perpendicular to the river bank, we have $C_1D_1//AE$ and $C_1D_1 = AE$. So, $AED_1C_1$ is a parallelogram (see Remark of Example 3 in Chapter 19) and $AC_1 = ED_1$. That is, the total length of the road is $ED_1 + D_1B$. This total length $\geqslant EB = ED + DB = AC + DB$. So, the above design of the road has the shortest total length.

5. (1) $AK$, $AM$, $AN$, $AH$, $AB$, $KM$, $KN$, $KH$, $KB$, $MN$, $MH$, $MB$, $NH$, $NB$, and $HB$.

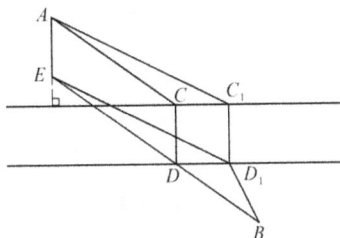

Fig. 9.3

(2) $M$ is the midpoint of $KN$ and $AH$.

(3) $N$ is a trisection point of $KH$ and $MB$.

6. $E$ is also the midpoint of $AD$. The reason is as follows: as $AB = CD$, $AB - CB = CD - CB$, which is

$$AC = BD. \tag{9.1}$$

Also, as $E$ is the midpoint of $BC$, we have

$$CE = EB. \tag{9.2}$$

On $(9.1) + (9.2)$, we have $AC + CE = EB + BD$, which is $AE = ED$. So, $E$ is the midpoint of $AD$.

7. $CF + FD + FE = CD + FD + DE$. As $C$, $D$, and $E$ are the quadrisection point of the line segment $AB$, $CD = DE = \frac{1}{4}AB$. As $F$ is a trisection point of the line segment $AB$, $FD = AD - AF = \frac{1}{2}AB - \frac{1}{3}AB = \frac{1}{6}AB$. Therefore, we have

$$CF + FD + FE = CD + FD + DE$$
$$= \frac{1}{4}AB + \frac{1}{6}AB + \frac{1}{4}AB$$
$$= \frac{2}{3}AB = \frac{2}{3} \times 12 = 8\text{(cm)}.$$

8. As $AC : CD : DB = 1 : 2 : 4$, $AC = \frac{1}{1+2+4}AB = \frac{1}{7}AB$, $CD = \frac{2}{7}AB$, and $DB = \frac{4}{7}AB$. Also, as $AM = \frac{1}{2}AC$, we have

$$MC = \frac{1}{2}AC = \frac{1}{2} \times \frac{1}{7}AB = \frac{1}{14}AB,$$

$$DN = \frac{1}{4}DB = \frac{1}{4} \times \frac{4}{7}AB = \frac{1}{7}AB.$$

Therefore, $MN = MC + CD + DN = \frac{1}{14}AB + \frac{2}{7}AB + \frac{1}{7}AB = \frac{1}{2}AB = \frac{1}{2} \times 14 = 7$.

9. As $AB = CD$ and $CB = \frac{1}{5}AB$, we have

$$AD = AB + CD - CB = 2AB - \frac{1}{5}AB = \frac{9}{5}AB.$$

Also, as $E$ and $F$ are the midpoints of $AB$ and $CD$, respectively, $AE = \frac{1}{2}AB$ and $FD = \frac{1}{2}CD = \frac{1}{2}AB$, we have

$$AD = AE + EF + FD = \frac{1}{2}AB + 12 + \frac{1}{2}AB = AB + 12.$$

Then, $\frac{9}{5}AB = AB + 12$, $\frac{4}{5}AB = 12$, $AB = 12 \times \frac{5}{4} = 15$(cm).

10. $AM + ND = AD - MN = a - b.$
$AB + CD = 2(AM + ND) = 2(a - b).$
$BC = AD - (AB + CD) = a - 2(a - b) = 2b - a.$

11. As $AP + PQ = 14 + 13 = 27 > 26 = AB$ together with $P$ and $Q$ are on the line segment $AB$, $Q$ must be between $A$ and $P$.

Fig. 9.4

$$BQ = AB - AQ = AB - (AP - PQ) = 26 - (14 - 13)$$
$$= 25\text{(cm)}.$$

12. As $D$ is the midpoint of $BC$, it is of course on the same straight line as $A$, $B$, and $C$. But there are two cases for the position of $D$.

(1) $D$ is on the same side of $A$ as $B$.

In this case (see Figure 9.5), as $AD = 12 < 16 = AB$, $D$ is on the line segment $AB$:

$$BC = 2BD = 2 \times (AB - AD) = 2 \times (16 - 12) = 8.$$

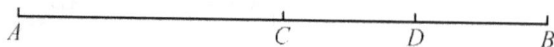

Fig. 9.5

(2) $D$ is on the other side of $A$ from $B$. In this case (see Figure 9.6),

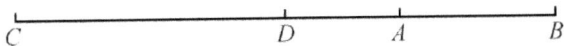

Fig. 9.6

$$DB = DA + AB = 12 + 16 = 28,$$
$$BC = 2BD = 2 \times 28 = 56.$$

# Solution 10

# Angles

1. (1) $170° - 20° = 150°$ is still an obtuse angle, so the proposition is false.
   (2) When subtracting an acute angle from a right angle, the difference is less than a right angle, so the difference must be an acute angle. Thus, the proposition is true.

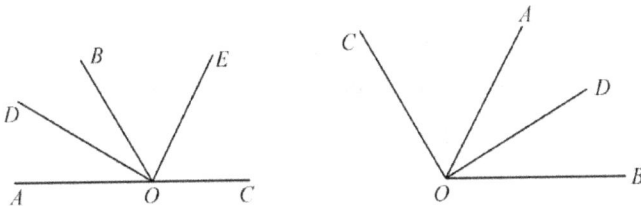

Fig. 10.1

   (3) As shown in the left figure of Figure 10.1, $\angle AOB$ and $\angle BOC$ are supplementary and their bisectors $OD$ and $OE$ form a right angle. But in the right figure, $\angle AOB = 60°$ and $\angle BOC = 120°$. $\angle AOB$ and $\angle BOC$ are supplementary and have a common side $OB$, but their bisectors $OD$ and $OA$ are not perpendicular ($\angle AOD = 30°$ is not a right angle). Thus, the proposition is false.
   (4) As shown in the right figure of Figure 10.1, $\angle AOB = \angle AOC$, but these two angles are not opposite, so the proposition is false.

(5) As shown in the right figure of Figure 10.1, $\angle AOB = \angle AOC$, and they have a common side $OA$, but the other sides $OB$ and $OC$ are not on a straight line, so the proposition is false.

2. Suppose this angle is $x°$, then its complementary angle is $(90-x)°$. So, $x - (90 - x) = 20$, $2x = 110$, $x = 55$, $90 - x = 35$, and $180 - 35 = 145$. The degree of the supplementary angle of the complementary angle of this angle is $145°$.

3. Suppose this angle is $x°$, then its complementary angle is $(90 - x)°$ and its supplementary angle is $(180 - x)°$. So, $(180 - x) + (90 - x) = (180 - x) - (90 - x) + 60$, $x = 60$, $90 - x = 30$, and the degree of the complementary angle of this angle is $30°$.

4.
$$\angle MON = \angle COM - \angle CON = \frac{1}{2}\angle BOC - \frac{1}{2}\angle AOC$$
$$= \frac{1}{2}(\angle BOC - \angle AOC) = \frac{1}{2} \times 90° = 45°.$$

5. $\angle AOE = \frac{1}{2}\angle AOB = \frac{1}{2}(90° - \angle BOC) = \frac{1}{2}(90° - 50°) = 20°$.
$\angle AOD = \angle AOB + \angle BOC + \angle COD = 40° + 50° + 40° = 130°$.
$\angle EOF = \angle AOD - \angle AOE - \angle DOF = 130° - 20° - 20° = 90°$.

6. Suppose this angle is $x°$, then $(180 - x) + 3(90 - x) = 360 \times \frac{11}{12}$, that is, $4x = 120$, $x = 30$. The degree of this angle is $30°$.

7. As $\angle BOC + \angle BOD = 180°$ and $\angle BOC - \angle BOD = 20°$, we have
$$\angle BOC = \frac{180° + 20°}{2} = 100°,$$
$$\angle BOD = \frac{180° - 20°}{2} = 80°.$$

As opposite angles are equal, we have $\angle AOC = \angle BOD = 80°$. As $OE$ bisects $\angle AOC$, we have $\angle COE = \frac{1}{2}\angle AOC = 40°$.
$$\angle BOE = \angle BOC + \angle COE = 100° + 40° = 140°.$$

8. (1) There are $\frac{5 \times 4}{2} = 10$ angles in total.
(2) There are $\frac{2000 \times 1999}{2} = 1999000$ angles in total.

9. As $OD$ is the bisector of $\angle AOC$, $\angle DOC = \frac{1}{2}\angle AOC$. For the same reason, we have $\angle COE = \frac{1}{2}\angle COB$. As $\angle AOC + \angle COB = \angle AOB = 180°$,

we have

$$\angle DOE = \angle DOC + \angle COE$$

$$= \frac{1}{2}\angle AOC + \frac{1}{2}\angle COB$$

$$= \frac{1}{2}(\angle AOC + \angle COB)$$

$$= \frac{1}{2} \times 180° = 90°.$$

10. Suppose this acute angle is $\alpha$, then the complementary angle of half of this angle is $90° - \frac{\alpha}{2}$ and twice its supplementary angle is $180° - 2\alpha$. Now, we have

$$2\left(90° - \frac{\alpha}{2}\right) - (180° - 2\alpha) = (180° - \alpha) - 180° + 2\alpha = \alpha.$$

11. $\triangle AOB$ and $\triangle OCD$ have the same shape obtained by folding along the diagonal of a $3 \times 1$ grid. If we cut out these two shapes and put them together, they coincide. Thus, $\angle 1$ and $\angle 9$ are complementary. For the same reason, $\angle 2$ and $\angle 6$ are complementary, and $\angle 4$ and $\angle 8$ are complementary. Furthermore, $\angle 3$, $\angle 5$, and $\angle 7$ are all $45°$. Thus, $\angle 1 + \angle 2 + \cdots + \angle 9 = (\angle 1 + \angle 9) + (\angle 2 + \angle 6) + (\angle 4 + \angle 8) + \angle 3 + \angle 5 + \angle 7 = 90° + 90° + 90° + 45° + 45° + 45° = 405°$.

12. Suppose $\angle A_2OA_1 = x°$, then the degrees of $\angle A_3OA_2$, $\angle A_4OA_3, \ldots$, $\angle A_8OA_7$, and $\angle A_1OA_8$ are $x + 4$, $x + 2 \times 4, \ldots, x + 6 \times 4$, $x + 7 \times 4$, respectively. The sum of them is $x + (x + 4) + (x + 2 \times 4) + \cdots + (x + 7 \times 4) = 8x + 4 \times (1 + 2 + \cdots + 7) = 8x + 4 \times \frac{7 \times 8}{2} = 8x + 112$. The sum of them is a round angle, thus $8x + 112 = 360$, $x = 31$. So, $\angle A_2OA_3 = \angle A_2OA_1 + 4° = 31° + 4° = 35°$.

13. At 8 o'clock, the minute hand is at 12 and the hour hand is at 8. The angle they form is $180° + 60° = 240°$. At $8:20$, the minute hand advances to 4 from 12 and turns $120°$. The hour hand advances from 8 and turns $120° \times \frac{1}{12} = 10°$. Thus, the angle formed by the minute hand and the hour hand is $240° - 120° + 10° = 130°$.

14. At 3 o'clock, the hour hand and the minute hand form a right angle. Suppose at $x$ minutes after three, they also form a right angle. At this time, the minute hand must be ahead of the hour hand (as the minute hand turns faster than the hour hand, the angle formed by the former is first decreasing, until the minute and hour hands coincide; after that, the angle formed by them is increasing until 90°). As 1 min mark on the clock is just 1 min or 6° ($360° \div 60 = 6°$), the minute hand turns $(6x)°$ and the hour hand turns $\frac{1}{12} \times (6x)° = (\frac{x}{2})°$. The angle formed by them is $(6x - 90 - \frac{x}{2})°$, and so $6x - 90 - \frac{x}{2} = 90$; on solving it, we have $x = \frac{360}{11} = 32\frac{8}{11}$. That is, at $32\frac{8}{11}$ min after 3, they form a right angle.

15. As shown in Figure 10.2, there are two cases.

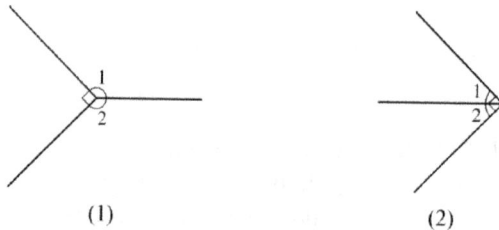

(1)                                    (2)

Fig. 10.2

(1) Both angles are $\frac{360° - 90°}{2} = 135°$.
(2) Both angles are $\frac{90°}{2} = 45°$.

# Solution 11

# Sum of the Interior Angles of a Triangle

1. The sum of the interior angles is $(10-2) \times 180° = 1440°$. The sum of the exterior angles is $360°$ (see Example 2).

2. The second interior angle is $56° \times 2 = 112°$. The third interior angle is $112° - 10° = 102°$. So, the fourth interior angle is $360° - 56° - 112° - 102° = 90°$.

3. Suppose $E$ is on the extension line of $BC$, then $\angle DCE = \angle DBC + \angle D$ and $\angle ACE = \angle ABC + \angle A$. Take the former formula subtracted by half of the latter formula, and combining the result with $\angle DCE = \frac{1}{2}\angle ACE$ and $\angle DBC = \frac{1}{2}\angle ABC$, we have $\angle D = \frac{1}{2}\angle A = \frac{1}{2} \times 80° = 40°$.

4. As shown in Figure 11.1, suppose $BD$ and $CE$ intersect at $F$. Extend $BD$ to intersect $AC$ at $G$. We have,

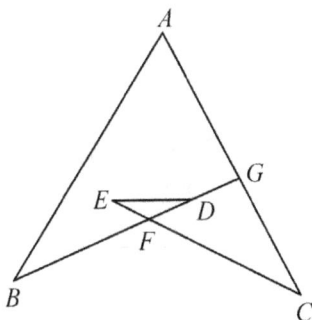

Fig. 11.1

$$\angle E + \angle D = 180° - \angle EFD$$

$$= 180° - \angle BFC,$$

$$\angle BFC = \angle C + \angle FGC$$

$$= \angle C + \angle A + \angle B.$$

So, $\angle A + \angle B + \angle C + \angle D + \angle E = 180°$.

5. As shown in Figure 11.2, suppose $AD$ and $CE$ intersect at $F$, and $AD$ and $BE$ intersect at $G$.

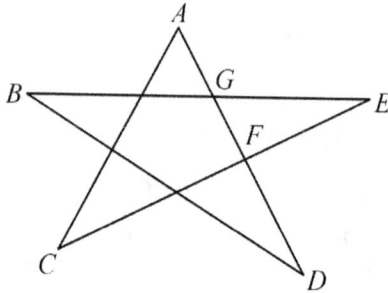

Fig. 11.2

We have

$$\angle A + \angle C = \angle CFD,$$

$$\angle DFE = \angle DGE + \angle E = \angle B + \angle D + \angle E.$$

So, $\angle A + \angle B + \angle C + \angle D + \angle E = \angle CFD + \angle DFE = 180°$.

6. As shown in Figure 11.3, connect $AD$. $\angle E + \angle F = 180° - \angle EGF = 180° - \angle AGD = \angle GAD + \angle GDA$. $\angle E + \angle F + \angle B + \angle C + \angle BAF + \angle CDE = (\angle GAD + \angle BAF) + (\angle GDA + \angle CDE) + \angle B + \angle C = \angle BAD + \angle CDA + \angle B + \angle C = 360°$. So, the answer is $360°$.

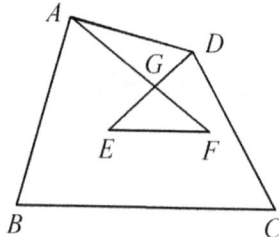

Fig. 11.3

7. $\angle ACB = \angle ADF + \angle CED = 100° + 35° = 135°$. $\angle B = 180° - \angle A -$
$\angle ACB = 180° - 20° - 135° = 25°$.
8. As $BD \perp AC$, $\angle BDC = 90°$,

$$\angle DBC = 90° - \angle C. \tag{11.1}$$

As $\angle ABC = \angle C$, we have

$$\angle A = 180° - \angle ABC - \angle C = 180° - 2\angle C,$$

$$\frac{1}{2}\angle A = 90° - \angle C. \tag{11.2}$$

By (11.1) and (11.2), we have $\angle DBC = \frac{1}{2}\angle A$.
9. As $AD$ is the bisector of $\angle BAC$, we have

$$\angle BAE = \angle DAC. \tag{11.3}$$

As $\angle ADB$ is an exterior angle of $\triangle BDE$, $\angle ADB > \angle BED = 90°$
and $\angle ACB = \angle ADB - \angle DAC > 90° - \angle DAC$. By (11.3), we have
$\angle ACB > 90° - \angle BAE = \angle ABE$.
10. $\angle BAC > \angle ACD$. As $CD$ is the bisector of an exterior angle of $\triangle ABC$,
$\angle ACD = \frac{1}{2}\angle ACE$, we have

$$\angle BAC > \frac{1}{2}\angle ACE. \tag{11.4}$$

Also,

$$\angle ACE = \angle B + \angle BAC. \tag{11.5}$$

By (11.4) and (11.5), we have $\angle BAC > \frac{1}{2}(\angle B + \angle BAC)$, that is,
$\angle BAC > \angle B$.
11. From Example 7, we have $\angle CEB = \angle F + \angle FCE + \angle FBE$. As $CF$
and $BF$ are the bisectors of $\angle ACD$ and $\angle ABD$, respectively, $\angle FCE =$
$\frac{1}{2}\angle ACE$ and $\angle FBE = \frac{1}{2}\angle DBE$. Now, it follows that

$$\angle CEB = \angle F + \frac{1}{2}(\angle ACE + \angle DBE). \tag{11.6}$$

Also, as $\angle CEB = \angle A + \angle ACE$ and $\angle CEB = \angle D + \angle DBE$, we have
$2\angle CEB = \angle A + \angle D + \angle ACE + \angle DBE$, that is,

$$\angle CEB = \frac{1}{2}(\angle A + \angle D) + \frac{1}{2}(\angle ACE + \angle DBE). \tag{11.7}$$

By (11.6) and (11.7), we have $\angle F = \frac{1}{2}(\angle A + \angle D)$.

12. From Example 7, we have

$$\angle BPC = \angle A + \angle PBA + \angle PCA > \angle A.$$

13. The sum of the interior angles of a pentagon is $(5-2) \times 180° = 3 \times 180°$. According to the given condition, the five interior angles form an arithmetic sequence, then the third angle is simply their average, and so the third angle is $3 \times 180° \div 5 = 108°$.

14. (1) If $3\angle A = 5\angle B$ and $3\angle C = 2\angle B$, then $\angle A = \frac{5}{3}\angle B$ and $\angle C = \frac{2}{3}\angle B$. By $\angle A + \angle B + \angle C = 180°$, we have $(\frac{5}{3} + 1 + \frac{2}{3})\angle B = 180°$, $\angle B = 54°$. So, $\angle A = \frac{5}{3}\angle B = 90°$, $\angle C = \frac{2}{3}\angle B = 36°$. Now, we set $\angle B = 54°$, $\angle A = 90° + \alpha$, and $\angle C = 36° - \alpha$, where $\alpha$ is an angle less than $36°$. For any $0° < \alpha < 36°$, $\angle A$, $\angle B$, and $\angle C$ can be the interior angles of a triangle (their sum is $180°$), and $3\angle A > 5\angle B$ and $3\angle C < 2\angle B$ satisfy the conditions.

   (2) As $3\angle A > 5\angle B = 3\angle B + 2\angle B \geqslant 3\angle B + 3\angle C$, $\angle A > \angle B + \angle C$. Thus, $\angle A$ is an obtuse angle, and $\triangle ABC$ is an obtuse triangle.

# Solution 12

# Parallel Lines

1. As shown in Figure 12.1, extend $AB$ to intersect $ED$ at $F$. As an exterior angle of $\triangle BEF$, $\angle ABE = \angle E + \angle BFE$. As $AB//CD$, $\angle BFE = \angle D$, and thus $\angle ABE = \angle D + \angle E$.

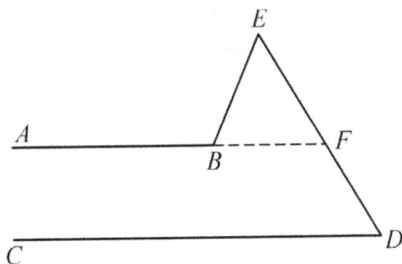

Fig. 12.1

2. As shown in Figure 12.2, construct parallel lines $EH$ and $GI$ of $AB$ through $E$ and $G$, respectively. As $EH//AB$, $\angle B = \angle 1$. As $EH//AB$ and $GI//AB$, $EH//GI$. By Example 3, we have $\angle F = \angle 2 + \angle 3$. As $GI//AB$ and $AB//CD$, $GI//CD$ and $\angle 4 = \angle D$. Adding the above three formulas, we have $\angle B + \angle F + \angle D = \angle 1 + \angle 2 + \angle 3 + \angle 4$, that is, $\angle B + \angle F + \angle D = \angle E + \angle G$.

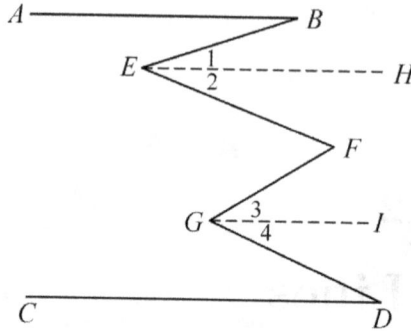

Fig. 12.2

3. As shown in Figure 12.3, construct $CF//AB$ through $C$. As $AB//CF$, $\angle ABC + \angle 1 = 180°$. As $\angle ABC + \angle BCD + \angle EDC = 360°$ and $\angle BCD = \angle 1 + \angle 2$, by subtraction, we have $\angle 2 + \angle EDC = 180°$, so $CF//ED$. As $CF//AB$ and $CF//ED$, $AB//ED$.

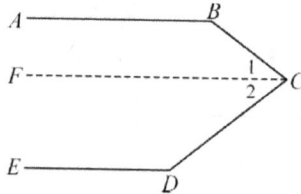

Fig. 12.3

4. (1) As $AB//CD$, $\angle A + \angle D = 180°$. As $\angle C = \angle A$, $\angle C + \angle D = 180°$, $AD//BC$.
   (2) As $AB//CD$, $\angle B + \angle C = 180°$. Since we have already proved that $\angle C + \angle D = 180°$ in (1), $\angle B = \angle D$.

5. As $AB//CD$, $\angle BEF + \angle EFD = 180°$. As $\angle EFD = 70°$, $\angle BEF = 180° - \angle EFD = 110°$. As $\angle 1 = \angle 2$, $\angle 2 = \frac{1}{2}\angle BEF = 55°$. As $AB//CD$, $\angle D = \angle 2 = 55°$.

6. $\angle ERQ = 180° - \angle FRG = 180° - 110° = 70°$. As $CD//EF$, $\angle SQP = \angle ERQ = 70°$. As $PS \perp GH$, $\angle SPQ = 90°$. In $\triangle PSQ$, we have
$$\angle PSQ = 180° - \angle SPQ - \angle SQP = 180° - 90° - 70° = 20°.$$

7. As $AB//CD$ and $CD//EF$, $AB//EF$. As $AB//EF$ and $\angle A = 105°$, $\angle CFE = \angle A = 105°$. In $\triangle CEF$, as $\angle ACE = 51°$, we have
$$\angle E = 180° - \angle ACE - \angle CFE = 180° - 51° - 105° = 24°.$$

8. As $BO$ and $CO$ are the bisectors of $\angle ABC$ and $\angle ACB$, respectively, $\angle OBC = \frac{1}{2}\angle ABC$ and $\angle OCB = \frac{1}{2}\angle ACB$. Therefore, $\angle OBC$ : $\angle OCB = \angle ABC : \angle ACB = 3 : 2$. Also, as $\angle OBC + \angle OCB = 180° - \angle BOC = 180° - 130° = 50°$, $\angle OBC = 50° \times \frac{3}{3+2} = 30°$ and $\angle OCB = 50° \times \frac{2}{3+2} = 20°$. Then, $\angle ABC = 2\angle OBC = 60°$ and $\angle ACB = 2\angle OCB = 40°$. As $EF//BC$, $\angle AEF = \angle ABC = 60°$ and $\angle EFC = 180° - \angle ACB = 140°$.

9. $\angle CBE = 180° - \angle ABE = \angle DEB$ and $\angle 1 = \angle 2$; taking the difference between these two formulas, we have $\angle FBE = \angle GEB$, and so $BF//GE$. As $BF//GE$, we have $\angle F = \angle G$.

10. As $DE$ bisects $\angle ADC$ and $CE$ bisects $\angle BCD$, we get $\angle ADC = 2\angle 1$ and $\angle BCD = 2\angle 2$. As $\angle 1 + \angle 2 = 90°$, $\angle ADC + \angle BCD = 2\angle 1 + 2\angle 2 = 2(\angle 1 + \angle 2) = 2 \times 90° = 180°$. As $\angle ADC + \angle BCD = 180°$, $AD//BC$. As $AD//BC$, $\angle A + \angle B = 180°$. As $DA \perp AB$, $\angle A = 90°$. As $\angle B = 180° - \angle A = 90°$, so $BC \perp AB$.

11. Take any point $O$. Through $O$, construct eight straight lines parallel to the given eight straight lines. These eight straight lines that we constructed divide the full round angle into 16 parts, and one of them must be $\leqslant \frac{360°}{16} < 23°$. This angle is equal to the angle formed by two of the given lines (these two given lines are parallel to the two sides of this angle, respectively). So, among all angles, there must be at least one angle that is less than $23°$.

# Solution 13

# Assigning Variables
# Unnecessary for Solution

1. Suppose the distance that Alice travels up the mountain is $s$ km, then the time required to travel up the mountain is $\frac{s}{4}$ h. The distance that she travels down the mountain is also $s$ km, and the time required to travel down the mountain is $\frac{s}{6}$ h. The average speed is

$$\frac{2s}{\frac{s}{4}+\frac{s}{6}} = \frac{2s}{\frac{5}{12}s} = \frac{24}{5} = 4.8(\text{km/h}).$$

2. Suppose the distance of a one-way trip is $s$ km, then

$$\text{the average speed} = \frac{2s}{\frac{s}{10}+\frac{s}{6}} = \frac{15}{2} = 7.5(\text{km/h}).$$

3. Suppose the speed of the ship is $x$ m/min and the speed of the stream is $y$ m/min. Then, the speed of the ship sailing downstream is $(x+y)$ m/min, and the speed of the ship sailing upstream is $(x-y)$ m/min. When they found that a wooden barrel had dropped, the distance between the barrel and the ship is $5[(x-y)+y]$ m. The time it took to catch up with the wooden barrel is

$$\frac{5[(x-y)+y]}{(x+y)-y} = 5(\text{min}).$$

4. The same amount of alcohol is in the two cups. Suppose that there is $V$ liters of water or alcohol in the cups originally, and there is $x$ liters of alcohol in cup $A$ and $y$ liters of water in cup $B$ at the end. When the whole process of pouring is complete, both cup $A$ and cup $B$ have $V$ liters of mixed liquid. Thus, there is $(V-y)$ liters of alcohol in cup $B$ and $x+(V-y)=V$, yielding $x=y$.

5. According to the problem, $a \times 3^5 + b \times 3^3 + c \times 3 + 3 = 48$. So, when $x = -3$, the value given by the formula is $a \times (-3)^5 + b \times (-3)^3 + c \times (-3) + 3 = -(a \times 3^5 + b \times 3^3 + c \times 3) + 3 = -45 + 3 = -42$.

6. Suppose that the workloads that $A$, $B$, $C$, and $D$ can handle in one day are $a$, $b$, $c$, and $d$, respectively. Then, we have

$$\begin{cases} a + b + c = \dfrac{1}{10}, & (13.1) \\[2mm] a + c + d = \dfrac{1}{8}, & (13.2) \\[2mm] b + d = \dfrac{1}{20}. & (13.3) \end{cases}$$

By $(13.1) - (13.2)$, we find

$$b - d = -\frac{1}{40}. \tag{13.4}$$

By $(13.3) - (13.4)$, we get $d = \frac{3}{80}$, that is, if $D$ works alone, it will take $\frac{80}{3}$ days to complete the project.

7. Suppose that the distance between stations $A$ and $B$ is $s$ m, then the distance between stations $B$ and $C$ is also $s$ m. Suppose the speed of Alice is $x$ m/s and the speed of Bob is $y$ m/s. According to the problem conditions, we have

$$\begin{cases} \dfrac{s + 100}{x} = \dfrac{s - 100}{y}, & (13.5) \\[4mm] \dfrac{(s - 100) + s + 300}{x} = \dfrac{100 + 300}{y}. & (13.6) \end{cases}$$

By $(13.6)$, we have $\frac{2s+200}{x} = \frac{400}{y}$, that is,

$$\frac{s + 100}{x} = \frac{200}{y}. \tag{13.7}$$

By $(13.5)$ and $(13.7)$, we find $\frac{s-100}{y} = \frac{200}{y}$, so $s - 100 = 200$, $s = 300$ (m).

**Another Solution**  Suppose that $s$ has the same meaning as above. When Alice travels $(s + 100)$ m, Bob travels $(s - 100)$ m; when Alice travels $(3s+300)$ m, Bob travels $(s+300)$ m. As $3s+300$ is three times $s + 100$, so $3(s - 100) = s + 300$; on solving it, we have $s = 300$ (m).

8. Suppose the dimensions of the rectangle with area 25 are $a$ and $b$. The dimensions of the rectangle with area 30 are $c$ and $d$. Then, the length and width of the rectangle with area 20 are $c$ and $b$, respectively, and the length and width of the shaded rectangle are $a$ and $d$, respectively. Therefore, the shaded area is

$$ad = \frac{(ab) \cdot (cd)}{(bc)} = \frac{25 \times 30}{20} = 37.5$$

9. Suppose that it takes $x$ yuan to buy one piece of $A$, $y$ yuan to buy one piece of $B$, and $z$ yuan to buy one piece of $C$. Then, we have

$$\begin{cases} 3x + 7y + z = 300, & (13.8) \\ 4x + 10y + z = 400. & (13.9) \end{cases}$$

On $(13.8) \times a + (13.9) \times b$, we have

$$(3a + 4b)x + (7a + 10b)y + (a + b)z = 300a + 400b. \qquad (13.10)$$

By assuming

$$3a + 4b = 7a + 10b = a + b, \qquad (13.11)$$

we have

$$2a + 3b = 0. \qquad (13.12)$$

We let $a = 3$ and $b = -2$, then (13.12) and (13.11) both hold, and (13.10) is simply $x + y + z = 100$, so it takes 100 yuan to buy one piece of each of $A$, $B$, and $C$.

Remark: If the coefficients of $x$, $y$, and $z$ on the left side of (13.10) are not 1 (but by (13.11), they can be equal), then we can find the value of $x + y + z$ by dividing both sides of (13.10) by $a + b$, which is simply the coefficient of $x$, $y$, or $z$. If the coefficients cannot be equal, then we need additional information to find the answer.

10. Suppose the original concentrations of alcohol in cup $A$ and cup $B$ are $a$ and $b$, respectively. The weight of the mixed solution $A_1$ poured out from cup $A$ is $x$ g. Given the problem conditions, we have

$$\frac{(16 - x)a + xb}{16} = \frac{(24 - x)b + xa}{24}.$$

After simplification, we have $48(a - b) = 5x(a - b)$; as $a \neq b$, we have $x = \frac{48}{5} = 9.6$ (g).

11. Suppose the person starts breakfast at $x$ min after 8 o'clock and finishes at $y$ min after 8 o'clock. At 8 o'clock, the angle between the hour hand and the minute hand is $240°$. The minute hand turns $\frac{360°}{60} = 6°$ every minute, and the hour hand turns $\frac{5}{60} \times 6° = \left(\frac{1}{2}\right)°$ every minute. According to the problem, we have

$$
\begin{cases}
6x - \dfrac{x}{2} = 240 - 25, & (13.13) \\[2mm]
6y - \dfrac{y}{2} = 240 + 25. & (13.14)
\end{cases}
$$

By $(13.14) - (13.13)$, we have $\frac{11}{2}(y - x) = 50$, that is, $y - x = \frac{100}{11}$. So, it takes this person $\frac{100}{11}$ min to have breakfast.

**Another Solution**   Suppose it takes this person $t$ minutes to have breakfast. As the minute hand is $25°$ behind the hour hand, the minute hand turns $(25 + 25)°$ more than the hour hand during the breakfast, $6(t - \frac{t}{12}) = 25 + 25$, and we have $t = \frac{100}{11}$.

12. Suppose that the amount of water originally in the reservoir is $a$, the amount of water flowing into the reservoir is $b$ per day, and the amount of water flowing out of the reservoir is $c$ per day. Furthermore, suppose that water runs out in $x$ days if the daily amount of water flowing out of the reservoir does not change. Then, we have

$$
\begin{cases}
a + 40b = 40c, & (13.15) \\
a + 40(1 + 0.2)b = 40(1 + 0.1)c, & (13.16) \\
a + (1 + 0.2)bx = cx. & (13.17)
\end{cases}
$$

On $(13.16) - (13.15)$, we find $c = 2b$. Substitute it in $(13.15)$, and we get $a = 40b$. Substitute $c = 2b$ and $a = 40b$ in $(13.17)$, and we have $40b + 1.2bx = 2bx$, that is, $0.8bx = 40b$, $x = 50$ (days).

13. Suppose that we take $x$, $y$, and $z$ g of the three mixtures for a total of $(x + y + z)$ g. According to the problem conditions, it contains $\left(\frac{3}{8}x + \frac{2}{5}z\right)$ g of $A$, $\left(\frac{5}{8}x + \frac{1}{3}y\right)$ g of $B$, and $\left(\frac{2}{3}y + \frac{3}{5}z\right)$ g of $C$. So, the ratio of ingredient $A$ is $\frac{\frac{3}{8}x + \frac{2}{5}z}{x + y + z} = \frac{3}{3 + 5 + 2}$, that is,

$$
\frac{3}{8}x + \frac{2}{5}z = \frac{3}{10}(x + y + z).
$$

For the same reason, the ratios of ingredients $B$ and $C$ can give two more formulas, and together, we have

$$\begin{cases} \dfrac{3}{8}x + \dfrac{2}{5}z = \dfrac{3}{10}(x+y+z), & (13.18) \\[2mm] \dfrac{5}{8}x + \dfrac{1}{3}y = \dfrac{5}{10}(x+y+z), & (13.19) \\[2mm] \dfrac{2}{3}y + \dfrac{3}{5}z = \dfrac{2}{10}(x+y+z). & (13.20) \end{cases}$$

On $(13.18) \times 5 - (13.19) \times 3$, we find

$$y = 2z. \tag{13.21}$$

Substituting it in $(13.18)$, we get

$$x = \frac{20}{3}z. \tag{13.22}$$

Therefore,

$$x : y : z = 20 : 6 : 3. \tag{13.23}$$

Remark: On $(13.18) + (13.19) + (13.20)$, we have $x + y + z = x+y+z$ or $0 = 0$. In other words, the values of $x$, $y$, and $z$ that satisfy $(13.23)$ also satisfy $(13.20)$. So, equation $(13.20)$ is not necessary.

14. Suppose that the age of the grandfather is $10a + b$ (where $a$ and $b$ are digits and $a \neq 0$), then the age of the father is $10b + a$. Given the problem conditions, the difference of the ages $(10a + b) - (10b + a) = 9(a - b)$ is a multiple of 4. As 4 and 9 are co-prime, $a - b$ must be a multiple of 4, that is, $a - b = 4$ or 8. However, by common sense, the age of the father cannot be in the teens, so $b \geqslant 2$, $a - b \leqslant 9 - 2 = 7$. Thus, $a - b = 4$. Bob's age is $9(a - b) \div 4 = 9 \times 4 \div 4 = 9$.

**Answer** Bob is 9 years old.

# Solution 14

# Undetermined Coefficients

1. $(x + 2)(x - 5) = x^2 - 3x - 10$, so $p = -3$ and $q = -10$.
2. According to the given condition, we have
$$\begin{cases} a + b = 11, \\ -a + b = -3. \end{cases}$$

   So, $b = \frac{11-3}{2} = 4$ and $a = 11 - 4 = 7$. When $x = 3$,
   $$ax + b = 3a + b = 3 \times 7 + 4 = 25.$$

3. According to the given condition, we have $q = 2$ and $1 + p + q = 9$.
   So, $p = 9 - 1 - q = 6$. The polynomial is $x^2 + 6x + 2$.
4. Suppose the polynomial is $ax^2 + bx + c$, then
$$\begin{cases} c = 1, \\ a + b + c = 4, \\ a - b + c = 2. \end{cases}$$

   On solving it, we have $a = 2$, $b = 1$, and $c = 1$, and so the polynomial is $2x^2 + x + 1$.
5. $(x - 1)^2(x + 1) = (x - 1)(x^2 - 1) = x^3 - x^2 - x + 1$, so $a = -1$, $b = -1$, and $c = 1$.
6. $(x^2 + mx + n)^2 = x^4 + 2mx^3 + (m^2 + 2n)x^2 + 2mnx + n^2$, so $2m = -6$, $m^2 + 2n = 13$, $2mn = -12$, and $k = n^2$. On solving them, we have $m = -3$, $n = 2$, and $k = 4$.
7. $(2x + y + k)(3x - 4y + l) = 6x^2 - 5xy - 4y^2 + (2l + 3k)x + (l - 4k)y + kl$, so $2l + 3k = -11$, $l - 4k = 22$, and $m = kl$. On solving them, we have $l = 2$, $k = -5$, and $m = -10$.

8. $3x^2 - 4x + 7 = 3(x+1-1)^2 - 4(x+1-1) + 7 = 3(x+1)^2 - 6(x+1) + 3 - 4(x+1) + 4 + 7 = 3(x+1)^2 - 10(x+1) + 14.$

9. $(x+2)(x+3) = x^2 + 5x + 6$. Multiply by $(x+2)(x+3)$ on both sides of

$$\frac{x+4}{x^2+5x+6} = \frac{A}{x+2} + \frac{B}{x+3},$$

and we have

$$x + 4 = A(x+3) + B(x+2) = (A+B)x + (3A+2B).$$

So, we have

$$\begin{cases} A + B = 1, \\ 3A + 2B = 4. \end{cases}$$

On solving it, we have $A = 2$ and $B = -1$.

10. In the expansion of $(3x^2 + ax + b)(2x^2 - 4x + 5)$, the coefficient of $x^3$ is $2a - 12$ and the coefficient of $x$ is $5a - 4b$. So,

$$\begin{cases} 2a - 12 = 0, \\ 5a - 4b = 0. \end{cases}$$

On solving it, we have $a = 6$ and $b = \frac{15}{2}$.

11. After eliminating the denominators, we have $x^2 = A(x-1)(x+1) + Bx(x+1) + Cx(x-1)$. Setting $x = 0$, we have $A = 0$. Let $x = 1$ to find $B = \frac{1}{2}$. Let $x = -1$ to get $C = \frac{1}{2}$. So, $A = 0$, $B = \frac{1}{2}$, and $C = \frac{1}{2}$.

12. After eliminating the denominators, we have

$$x^2 - 2x + 5 = A(x-2)(x^2+1) + B(x^2+1) + (Cx+D)(x-2)^2.$$

Let $x = 2$. We have $B = 1$. So,

$$-2x + 4 = A(x-2)(x^2+1) + (Cx+D)(x-2)^2. \qquad (14.1)$$

Dividing by $x - 2$ on both sides, we have

$$-2 = A(x^2+1) + (Cx+D)(x-2). \qquad (14.2)$$

Let $x = 2$ to get $A = -\frac{2}{5}$. After comparing the coefficients of $x^2$ and constant terms on both sides of (14.2), we have $A + C = 0$ and $A - 2D = -2$, so $C = -A = \frac{2}{5}$ and $D = \frac{A+2}{2} = \frac{4}{5}$. So, $A = -\frac{2}{5}$, $B = 1$, $C = \frac{2}{5}$, and $D = \frac{4}{5}$.

Remark: Under the condition of $x \neq 2$, we have (14.2) from (14.1). But since (14.2) holds for all $x \neq 2$, it holds for all $x$ (including $x = 2$). Thus, we can set $x = 2$ to find the value of $A$.

13. Let $x = -5$, and we have $(5^2 - 5a + b)^{10} = -(-5c + d)^{20}$; therefore, we have $5^2 - 5a + b = -5c + d = 0$,

$$d = 5c. \tag{14.3}$$

Substituting (14.3) into the original formula, we have $(x^2 + ax + b)^{10} \equiv (x + 5)^{20} - c^{20}(x + 5)^{20} \equiv (1 - c^{20})(x + 5)^{20}$. After comparing the coefficients of the highest degree term $x^{20}$, we have $1 = 1 - c^{20}$, so $c = 0$, $d = 5c = 0$. The original formula becomes $(x^2 + ax + b)^{10} \equiv (x + 5)^{20} \equiv (x^2 + 10x + 25)^{10}$; therefore, $a = 10$ and $b = 25$. So, $a = 10$, $b = 25$, and $c = d = 0$.

14. Suppose $ax^3 + bx^2 + cx + d = (x^2 + p)(ax + q)$. Then,

$$ax^3 + bx^2 + cx + d = ax^3 + qx^2 + apx + pq.$$

After comparing the coefficients on both sides, we find

$$\begin{cases} b = q, \\ c = ap, \\ d = pq. \end{cases}$$

So, $ad = apq = bc$.

15. Suppose

$$x^3 + bx^2 + cx + d = (x^2 + px + q)(x + m), \tag{14.4}$$

where $p$, $q$, and $m$ are all integers. As $bd + cd = d(b + c)$ is odd, $d$ is odd and $b + c$ is also odd. After comparing the coefficients on both sides of (14.4), we have $d = qm$. As $d$ is odd, $m$ is odd. Let $x = 1$ in (14.4); we get

$$1 + (b + c) + d = (1 + p + q)(1 + m). \tag{14.5}$$

On the left side of 14.5, 1, $b + c$, and $d$ are all odd, so their sum is odd. So, the left side of 14.5 is odd. On the right side of 14.5, $1 + m$ is even. So, the right side of 14.5 is even. Thus, 14.5 does not hold. This contradiction shows that $x^3 + bx^2 + cx + d$ cannot be expressed as a product of two polynomials with integer coefficients.

## Solution 15

# Synthetic Division and Polynomial Remainder Theorem

1.
$$\begin{array}{r|rrrr} 2 & 3 & +0 & -5 & +6 \\ & & & & \\ \hline & 3 & +6 & +7 & +20 \end{array}$$

The quotient is $3x^2 + 6x + 7$, and the remainder is 20.

2.
$$\begin{array}{r|rrrrr} -3 & -4 & +2 & +5 & +0 & +8 \\ & & & & & \\ \hline & -4 & +14 & -37 & +111 & -325 \end{array}$$

The quotient is $-4x^3 + 14x^2 - 37x + 111$, and the remainder is $-325$.

3.
$$\begin{array}{r|rrrrr} \frac{1}{2} & -6 & +0 & -7 & +8 & +9 \\ & & & & & \\ \hline & -6 & -3 & -\frac{17}{2} & +\frac{15}{4} & +\frac{87}{8} \end{array}$$

The quotient is $-3x^3 - \frac{3}{2}x^2 - \frac{17}{4}x + \frac{15}{8}$, and the remainder is $\frac{87}{8}$.

4.
$$\begin{array}{r|rrrr} \frac{2}{3} & 27 & -9 & +5 & -2 \\ & & & & \\ \hline & 27 & +9 & +11 & +\frac{16}{3} \end{array}$$

The quotient is $9x^2 + 3x + \frac{11}{3}$, and the remainder is $\frac{16}{3}$.

5. $6 \times (-1)^5 - 4 \times (-1)^3 + 5 \times (-1)^2 + 3 \times (-1) + 8 = -6 + 4 + 5 - 3 + 8 = 8$.
   The remainder is 8.

6. $f(-3) = 0$, that is, $3^4 - 3 \times 3^3 + 8 \times 3^2 + 3k + 11 = 0$, $k = -\frac{83}{3}$.

7. $f(-\frac{1}{4}) = 0$, that is, $-2 \times (\frac{1}{4})^3 + (\frac{1}{4})^2 - \frac{1}{4}k - 2 = 0$, $k = -7\frac{7}{8}$.

8.

$$
\begin{array}{c|cccccc}
-\frac{1}{3} & 3 & -17 & +12 & +6 & +9 & +8 \\
\hline
 & 3 & -18 & +18 & +0 & +9 & +5
\end{array}
$$

By the remainder theorem, we have $f(-\frac{1}{3}) = 5$.

9.
$$
\begin{cases}
f(-1) = (-1)^4 - a \times (-1)^2 - b \times (-1) + 2 = -a + b + 3 = 0, \\
f(-2) = (-2)^4 - a \times (-2)^2 - b \times (-2) + 2 = -4a + 2b + 18 = 0.
\end{cases}
$$

On solving it, we have $a = 6$ and $b = 3$.

10. The remainder we need to find is equal to the remainder of $f(x)$ divided by $x - 2$, that is, $f(2) = 3 \times 2^4 - 8 \times 2^3 + 5 \times 2^5 - 2 + 8 = 150$.

11.
$$
\begin{cases}
f(-1) = 2 \times (-1)^3 - 3 \times (-1)^2 + a \times (-1) + b = -5 - a + b = 7, \\
f(1) = 2 \times 1^3 - 3 \times 1^2 + a \times 1 + b = -1 + a + b = 5.
\end{cases}
$$

On solving it, we have $a = -3$ and $b = 9$.

12. As $f(x)$ is a factor of $3(x^4 + 6x^2 + 25) - (3x^4 + 4x^2 + 28x + 5) = 14(x^2 - 2x + 5)$, $f(x) = x^2 - 2x + 5$.

13. Suppose that $f(x) = (x-1)(x-2)(x-3)q(x) + (ax^2 + bx + c)$, then we have

$$
\begin{cases}
f(1) = a + b + c = 1, \\
f(2) = 4a + 2b + c = 2, \\
f(3) = 9a + 3b + c = 3.
\end{cases}
$$

On solving it, we have

$$
\begin{cases}
a = 0, \\
b = 1, \\
c = 0.
\end{cases}
$$

So, the remainder is $x$.

**Another Solution**  The remainder of $f(x) - x$ divided by $x - 1$ is $f(1) - 1 = 0$, so $f(x) - x$ is divisible by $x - 1$. For the same reason $f(x) - x$ is divisible by $x - 2$ and $x - 3$. So, $f(x) - x$ is divisible by $(x-1)(x-2)(x-3)$. That is, the remainder of $f(x)$ divided by $(x-1)(x-2)(x-3)$ is $x$.

14.
$$\begin{cases} f(2) = 8a + 4b - 16 - 12 = 0, \\ f(3) = 27a + 9b - 24 - 12 = 0. \end{cases}$$

On solving it, we have

$$\begin{cases} a = -3, \\ b = 13. \end{cases}$$

| 2 | −3 | +13 | −8 | −12 |
|---|----|-----|-----|-----|
| 3 | −3 | +7 | +6 | |
| | −3 | −2 | | |

The quotient is $-3x - 2$.

15.

| 2 | 1 | +0 | +0 | +0 | −5q | +4r |
|---|---|-----|------|------|-------------|----------------------|
| 2 | 1 | +2 | +4 | +8 | +(16 − 5q) | +(32 − 10q + 4r) |
| | 1 | +4 | +12 | +32 | +(80 − 5q) | |

According to the problem conditions, we have

$$\begin{cases} 80 - 5q = 0, \\ 32 - 10q + 4r = 0, \end{cases}$$

On solving it, we have

$$\begin{cases} q = 16, \\ r = 32. \end{cases}$$

16. Suppose that $f(x) = ax^3 + bx^2 + cx + d(a \neq 0)$. As given in the problem, we have

$$\begin{cases} f(x) = (x^2 - 1)(ax + n) + 2x - 5, & (15.1) \\ f(x) = (x^2 - 4)(ax + m) - 3x + 4. & (15.2) \end{cases}$$

In (15.1), let $x = \pm 1$, and we get

$$\begin{cases} a + b + c + d = -3, & (15.3) \\ -a + b - c + d = -7. & (15.4) \end{cases}$$

In (15.2), let $x = \pm 2$, and we find

$$\begin{cases} 8a + 4b + 2c + d = -2, & (15.5) \\ -8a + 4b - 2c + d = 10. & (15.6) \end{cases}$$

By (15.3), (15.4), (15.5), and (15.6), we have $a = -\frac{5}{3}$, $b = 3$, $c = \frac{11}{3}$, and $d = -8$. So, the polynomial is $-\frac{5}{3}x^3 + 3x^2 + \frac{11}{3}x - 8$.

17. As given in the problem, we have $f(x) = a(x - a_1)(x - a_2)(x - a_3) + 1$ (where $a$ is an integer and $a \neq 0$). As $b$ is not equal to any of $a_1$, $a_2$, or $a_3$, we have $(b - a_1)(b - a_2)(b - a_3) \neq 0$,

$$f(b) = a(b - a_1)(b - a_2)(b - a_3) + 1 \neq 1.$$

# Solution 16

# Simplifying and Evaluating an Algebraic Formula

1. (1) $2^2 - \frac{1^2}{2} = 3\frac{1}{2}$.

   (2) $(\frac{2}{3})^2 - (\frac{1}{9})^2 \times \frac{3}{2} = \frac{4}{9} - \frac{1}{54} = \frac{23}{54}$.

2. $x = -(2 \times 3 \times \frac{1}{6})^3 = -1$.

   $x^{2007} + x^{2006} + \cdots + x + 1 = (-1) + 1 + \cdots + (-1) + 1 = 0$.

3. $a \times 3^3 + b \times 3 + 8 = 12$, so $a \times 3^3 + b \times 3 - 5 = (12 - 8) - 5 = -1$.

4. Suppose that $x = 3k$, $y = 4k$ and $z = 5k$, then $12k - 20k + 10k = 10$, $2k = 10$, and $k = 5$. Therefore, $x = 15$, $y = 20$ and $z = 25$. Then, $2x - 5y + z = 30 - 100 + 25 = -45$.

5. $a^2 + b^2 = (a + b)^2 - 2ab = 3^2 - 2 \times 2 = 5$.

6.
$$x^3 + y^3 + 3xy = (x + y)(x^2 - xy + y^2) + 3xy$$
$$= x^2 - xy + y^2 + 3xy \quad (\text{as } x + y = 1)$$
$$= x^2 + 2xy + y^2$$
$$= (x + y)^2$$
$$= 1.$$

7.
$$6x^3 + 7x^2 - 5x + 2006 = 2x(3x^2 - x - 1) + 9x^2 - 3x + 2006$$
$$= 2x(3x^2 - x - 1) + 3(3x^2 - x - 1) + 2009$$
$$= 2009 \quad (\text{as } 3x^2 - x - 1 = 0).$$

8. By $x^2 - 3x + 1 = 0$, we have $x \neq 0$. Therefore, divide $x$ on both sides, and we have

$$x + \frac{1}{x} = 3.$$

(1) $x^2 + \frac{1}{x^2} = (x + \frac{1}{x})^2 - 2 = 3^2 - 2 = 7.$

(2) $x^3 + \frac{1}{x^3} = 3(x^2 - 1 + \frac{1}{x^2}) = 3(7 - 1) = 18.$

(3) $x^4 + \frac{1}{x^4} = (x^2 + \frac{1}{x^2})^2 - 2 = 7^2 - 2 = 47.$

9. $a \times (-5)^4 + b \times (-5)^2 + c = a \times 5^4 + b \times 5^2 + c$. So, when $x = 5$ and $x = -5$, the value of $y$ is equal, which is 3.

10. (1) Let $x = 1$ in

$$(2x - 1)^5 = a_5 x^5 + a_4 x^4 + a_3 x^3 + a_2 x^2 + a_1 x + a_0. \qquad (16.1)$$

We have

$$a_0 + a_1 + a_2 + a_3 + a_4 + a_5 = 1. \qquad (16.2)$$

(2) Let $x = -1$ in (16.1), and we have

$$a_0 - a_1 + a_2 - a_3 + a_4 - a_5 = -243. \qquad (16.3)$$

(3) On (16.2) + (16.3), we have $2(a_0 + a_2 + a_4) = -242$, so $a_0 + a_2 + a_4 = -121.$

# Solution 17

# Logical Inference I

1. As 30 students participated in the sports team and 25 students participated in the art team, totally there are

$$30 + 25 = 55$$

participants.

As 13 students participated in both teams, they have been computed twice. As each student participated in at least one team, there are

$$55 - 13 = 42$$

students in the class.

2. If this sentence is true, then following what he has said, this sentence should be false. So, this sentence is false.

3. By condition (2), they will go to at least one of cities $D$ and $E$. If they go to city $E$, then by condition (5), they must go to cities $A$ and $D$. Then, by conditions (1) and (4), they must go to cities $B$ and $C$, which contradicts condition (3). So, they cannot go to city $E$, and they must go to city $D$. By condition (4), they must go to city $C$. In addition, by condition (3), they cannot go to city $B$; therefore, by condition (1), they cannot go to city $A$ either. So, the visiting group can at most go to cities $C$ and $D$. After checking, going to cities $C$ and $D$ satisfies all the conditions.

4. The inference process is as follows:

   (1) By ③, teacher Zhang does not teach biology, and by ④ and ②, teacher Li does not teach biology, so teacher Wang teaches biology.

   (2) By ⑤, teacher Li does not teach mathematics, and by ④ and (1), teacher Wang does not teach mathematics, so teacher Zhang teaches mathematics.

(3) By ⑤ and (2), teacher Wang teaches English.

(4) By (1) and (3), teacher Wang does not teach physical education. By ③, teacher Zhang does not teach physical education. So, teacher Li teaches physical education.

(5) By ①, teacher Li does not teach physics, so teacher Zhang teaches physics. The last subject history should be taught by teacher Li.

So, teacher Li teaches history and physical education; teacher Wang teaches English and biology; and teacher Zhang teaches mathematics and physics.

5. If what $A$ said is true, then what $B$ said is false and what $D$ said is false. Therefore, we infer that $D$ did not solve the problem first by what $B$ said; however, we also infer that $D$ solved the problem first by what $D$ said, resulting in a contradiction. Therefore, what $A$ said is false, and $A$ solved the problem first (now, only what $D$ said is true).

6. By (3) and (6), the second book $D$ read is not the first book $A$ or $C$ read. Note that the last books they read are just the five books they borrowed. As the second book $B$ read is the fifth book $A$ read, the fourth book $B$ read is the fifth book $C$ read, and the third book $C$ read is the fifth book $D$ read, so the third book $B$ read is not the fifth book $A$, $C$, or $D$ read. Then, it must be the fifth book $E$ read. So, the first book $B$ read is the fifth book $D$ read. Then, the second book $D$ read is not the first book $B$ read. Therefore, the second book $D$ read is the first book $E$ read.

7. We first put 9 pearls on each side of the balance. If they weigh the same, then the fake one must be among the remaining 9 pearls. If the two sides are unbalanced, then the fake one must be on the lighter side. So far, we have only weighed one time to determine which of the 9 pearls is the fake one. Now, put 3 pearls from these 9 pearls on each side of the balance to weigh the second time. Similarly, we can determine which of the 3 pearls is the fake one. Finally, we put 1 pearl on each of the balance and weigh for the third time. If they weigh the same, then the fake one is the remaining one. If it is unbalanced, then the lighter one is the fake one.

8. If Wang did the bad thing, then what Chen said are both true, which contradicts the given condition. So, Wang did not do it. If Zhang did the bad thing, then what Zhang said are both false, which contradicts the given condition. So, Zhang did not do it. Then, it must be Chen who did it. With this results, each of them told a lie and a truth.

9. (1) In a single round-robin match, there are totally 6 games. In each game, the two teams will get at most 3 points. So, the total number of points is at most 18.

   If a team gets 7 points, then the remaining teams get at most 11. It's impossible for another two teams to get at least 7 points each. (Actually, no other team can get more than 7 points at this time. Think about the reason.)

   Therefore, getting 7 points can ensure qualifying.

   On the other hand, if a team only gets 6 points, it cannot ensure qualifying. For example, all of $A$, $B$, and $C$ beat $D$. $A$ beats $B$, $B$ beats $C$, and $C$ beats $A$. Then $A$, $B$, and $C$ each gets 6 points, and now the first and the second will be determined according to the goal differences and the numbers of goals scored.

   So, to ensure qualifying, at least 7 points are necessary.

   (2) Getting 3 points may lead to qualification. If all 6 games are ties, then these 4 teams all get 3 points, and we need to compare the numbers of goals scored (as all games are ties, so the goal difference for each team is 0). If a team has scored a large number of goals, it may qualify.

   Getting 2 points may also lead to qualification. Suppose that $D$ beats $A$, $B$, and $C$. But games among $A$, $B$, and $C$ are ties. Then, $A$, $B$ and $C$ all get 2 points, and one of them will qualify.

   (3) According to the problem, this good thing can only be done by one of Hong, Qiang, and Hua. Therefore, we can use the assumption method.

   (1) If Hong did it, then what Hong said is false, what Qiang said is true, and what Hua said is true, which contradicts the assumption that two of them told lies and one told the truth. So, Hong did not do it.

   (2) If Qiang did it, then what Hong said is true and what Hua said is true, which contradicts that two of them told lies. So, Qiang did not do it.

   (3) If Hua did it, then what Hua said is false and what Hong said is false, but what Qiang said is true, which satisfies the given condition.

   So, Hua did this good thing.

# Solution 18

# Logical Inference II

1. As what $B$ said contradicts what $D$ said, we can judge that one of them told the truth and the other told a lie. As there is only one truth, what $C$ said must be false.

   So, $C$ did the bad thing. Then, $A$, $B$, and $C$ lied and $D$ told the truth.

2. The opposite letter of $a$ cannot be $d$ or $f$ and cannot be $b$ or $c$ either, so the opposite letter of $a$ must be $e$.

   For the same reason, the opposite letter of $d$ must be $b$; therefore, the opposite letter of $c$ must be $f$.

   So, the opposite letters of $a, b, c, d, e$, and $f$ are $e, d, f, b, a$, and $c$, respectively.

3. We draw a table as follows: the first line represents the colors of the beads, the first column represents these five individuals, and the numbers in the table represent which bags they guess.

   |   | Red | Blue | Yellow | White | Purple |
   |---|-----|------|--------|-------|--------|
   | $A$ |   |   | 3 |   | 2 |
   | $B$ | 4 | 2 |   |   |   |
   | $C$ | 1 |   |   | 5 |   |
   | $D$ |   | 3 |   | 4 |   |
   | $E$ |   |   | 2 |   | 5 |

As each bag has one person guessed correctly and number 1 only appears in the second row and fourth column, the bead in the first bag is red and $C$ guessed this correctly. Then, it is incorrect that $B$ guessed that the bead in the fourth bag is red, so the bead in the fourth bag is white and $D$ guessed this correctly. Then, the bead in the fifth bag is not white, and it should be purple, which $E$ guessed correctly. Then, $B$ guessed correctly that the bead in the second bag is blue, and $A$ guessed correctly that the bead in the third bag is yellow.

Finally, $A$ guessed correctly that the bead in the third bag is yellow, $B$ guessed correctly that the bead in the second bag is blue, $C$ guessed correctly that the bead in the first bag is red, $D$ guessed correctly that the bead in the fourth bag is white, and $E$ guessed correctly that the bead in the fifth bag is purple.

4. By conditions (1), (2), and (3), we infer that the three passengers live in Detroit, Chicago, and somewhere between Detroit and Chicago, respectively.

   By condition (6), $S$ is not the stoker.

   By conditions (1) and (3), $R$ is not the brakeman; if $R$ is the brakeman, then Mr. $R$ lives in Chicago, which contradicts condition (1).

   By conditions (4), (5), and $3 \nmid 20000$, Mr. $J$ is not the neighbor of the brakeman. By condition (2), Mr. $J$ lives in Detroit or Chicago.

   By condition (1), we have that Mr. $J$ did not live in Detroit.

   Therefore, Mr. $J$ lives in Chicago.

   By condition (3), $J$ is the brakeman.

   Therefore, $S$ is the driver.

   We conclude that $S$ is the driver, $J$ is the brakeman, and $R$ is the stoker. Mr. $S$ lives somewhere between Chicago and Detroit, Mr. $J$ lives in Chicago, and Mr. $R$ lives in Detroit.

5. The numbers of students who do not participate in Chinese, Mathematics, English, and Science courses are 7, 5, 4, and 6, respectively. So, the number of students not enrolled in at least one course is at most $7+5+4+6 = 22$. That is, there are at least $25 - 22 = 3$ students who participate in all four courses.

   We show an example where the number of students who participate in all four courses is exactly 3 as follows:

   There are 3 students participating in all four courses, 7 students participating in three courses except Chinese, 5 students participating in three courses except Mathematics, 4 students participating in three courses except English, and 6 students participating in three courses except Science. The total number is 25, which meets the requirements of this problem.

6. We use the backward method and list the numbers in the following table:

|  | $A$ | $B$ | $C$ |
|---|---|---|---|
| After the third time | 8 | 8 | 8 |
| After the second time | 4 | 4 | 16 |
| After the first time | 2 | 14 | 8 |
| Original numbers | 13 | 7 | 4 |

7. If there are only two students who had exchanged opinions with 4 students, then all the remaining students had only exchanged opinions with 3 students, and thus the number of times of exchanging opinions would be

$$11 \times 3 + 2 \times (4 - 3) = 35.$$

But this number should be an even number (if $A$ exchanged opinions with $B$, then $B$ exchanged opinions with $A$, and this is counted twice in the total number. So, the number of times of exchanging opinions should be an even number), which contradicts that 35 is an odd number. Therefore, there must be another student who exchanged opinions with at least 4 students.

8. Suppose that $A$ is the player who won the most times (maybe more than one player won the most times).

As $A$ fails to beat all the opponents, it must be $C$ who beats $A$. Among all the players who $A$ beats, it must be $B$ who beats $C$ (otherwise, $C$ wins at least one more time than $A$). Then, $A$ beats $B$, $B$ beats $C$, and $C$ beats $A$.

9. By condition (1), Li is not the cross-talk actor. By condition (3), Li was not born in Beijing. By condition (4), Li was not born in Wuhan. Therefore, Li was born in Shanghai. By conditions (2) and (1), Li is the dancer.

By condition (1), Wang is the cross-talk actor. By condition (3), Wang was born in Beijing.

Therefore, Li was born in Shanghai and is the dancer. Wang was born in Beijing and is the cross-talk actor. Zhang was born in Wuhan and is the singer.

10. We start backtracking from $(17, 1983, 1999)$. Suppose the last three numbers are $(17, 1983, x)$, $x + 17 - 1 = 1983$, so $x = 1983 + 1 - 17 = 1967$.

     Continuing in the same manner, we have

     $(17, 1983, 1999) \leftarrow (17, 1967, 1983) \leftarrow (17, 1951, 1967) \leftarrow \cdots \leftarrow$
     $(17, 1983 - 16 \times 123, 1983 - 16 \times 122)$. Then, $(17, 15, 31) \leftarrow (17, 15, 17 - 15 + 1)$, and $(17, 15, 3) \leftarrow (13, 15, 3) \leftarrow (13, 11, 3) \leftarrow (9, 11, 3) \leftarrow (9, 7, 3) \leftarrow (5, 7, 3) \leftarrow (5, 3, 3) \leftarrow (3, 3, 3)$.

     So, the original three numbers are all 3.

# Solution 19

# Divisibility

1. $3 \times 4 \times 5 = 60$. The numbers that are divisible by 3, 4, and 5 are exactly the numbers that are divisible by 60. $200 \div 60 = 3 \cdots 20$, so within 200, there are 3 positive integers that are divisible by 60. Thus, there are three positive integers that are divisible by 3, 4, and 5 (which are 60, 120, and 180).

2. The numbers that are divisible by 2, 4, 6, and 8 are exactly the numbers that are divisible by 24. As $1000 \div 24 = 41 \cdots 16$, within 1000 there are 41 numbers that are divisible by 2, 4, 6, and 8.

3. $88 = 8 \times 11$. The numbers that are divisible by 88 are the numbers that are divisible by 8 and 11. By $8 \mid \overline{3a123b}$, we have $8 \mid \overline{23b}$, so $8 \mid \overline{7b}$, $b = 2$. By $11 \mid \overline{3a1232}$, we have $11 \mid (3 + 1 + 3 - a - 2 - 2)$, that is, $11 \mid a - 3$. As $0 \leqslant a \leqslant 9$, $a = 3$.

4. As 11 and 13 are co-prime, the numbers that are divisible by 11 and 13 are the numbers that are divisible by $11 \times 13 = 143$. Since the quotient of $9999 \div 143$ is 69, the required largest four-digit number should be $143 \times 69 = 9867$.

5. Since $\overline{72xy}$ is divisible by 4 and 5, we have $y = 0$ and $x$ is even. Since $\overline{72x0}$ is divisible by 9, we have that $x$ is divisible by 9, so the even number $x = 0$. We can check that 7200 is divisible by 2, 3, 4, 5, 6, and 9. Then, $x = y = 0$.

6. $2 \times 3 \times 5 \times 11 = 330$. The numbers that are divisible by 2, 3, 5, and 11 are just the numbers that are divisible by 330. $1992400 \div 330 = 6037 \cdots 190$, so the smallest seven-digit number (whose first four digits are 1992) is $1992400 - 190 = 1992210$. That is, the answer to this problem is 1992210.

**Another Solution**   As the seven-digit number is divisible by 2 and 5, its ones digit is 0. As $\overline{1992xy0}$ is divisible by 11, $(1+9+x)-(9+2+y)$ is divisible by 11, that is, $x-y-1$ is divisible by 11. As $0 \leqslant x, y \leqslant 9$, we have

$$x = y + 1. \tag{19.1}$$

As $\overline{1992xy0}$ is divisible by 3, its digit sum is divisible by 3, that is, $x+y$ is a multiple of 3. By (19.1), we have that $2y + 1$ is a multiple of 3, so $y$ is at least 1 and $x$ is at least $1 + 1 = 2$. So, the smallest number is 1992210.

7. Let $a = 2 \times 3 \times 4 \times \cdots \times 17 \times 18$. Then, $a+2, a+3, \ldots, a+17$ and $a+18$ are divisible by 2, 3, \ldots, 17 and 18, respectively, and are greater than 2, 3, \ldots, 17 and 18, respectively. So, they are 17 consecutive composite numbers.

8. Each fountain pen costs 4 yuan and each ballpoint pen costs 0.8 yuan, so the total cost of the fountain and ballpoints pens should be a multiple of 0.04 yuan. As Bob bought 8 pencils and 12 erasers, their total cost should be a multiple of 0.04 yuan. So, the grand total cost should be a multiple of 0.04 yuan, but 20.1 yuan is not a multiple of 0.04 yuan. So, the salesperson miscalculated.

9. This three-digit number is divisible by 7 after subtracted by 7. So, this three-digit number is divisible by 7. For the same reason, this three-digit number is divisible by 8 and 9. Therefore, this three-digit number is divisible by $7 \times 8 \times 9 = 504$. As $504 \times 2 \geqslant 1000$, there is only one three-digit number, which is 504, that is divisible by 504. So, the three-digit number is 504.

10. The remainder of this two-digit number divided by 3 is 1, so this two-digit number is divisible by 3 after subtracting 1. For the same reason, this number is divisible by 4 and 5 after subtracting 1. Therefore, this two-digit number is divisible by $3 \times 4 \times 5 = 60$ after subtracting 1. As $60 \times 2 \geqslant 100$, there is only one two-digit number, which is 60, that is divisible by 60. So, the two-digit number is 61.

11. Within 100, there are 33 numbers divisible by 3, 25 numbers divisible by 4, and 8 numbers divisible by 12 ($100 \div 3 = 33 \cdots 1$, $100 \div 4 = 25$, $100 \div 12 = 8 \cdots 4$). Within 100, after eliminating the multiples of 3 or 4, the remaining numbers are not divisible by 3 or 4. But we need to pay attention to the numbers that are divisible by 12, as they are divisible both by 3 and 4. In the formula $100-33-25$, these numbers are counted twice. So, we must add 1 back for each such number. Therefore, the

number of positive integers within 100 that are neither divisible by 3 nor divisible by 4 is $100 - 33 - 25 + 8 = 50$.

12. As this four-digit number is divisible by 5, its ones digit must be 5. As this four-digit number is divisible by 3, its digit sum must be divisible by 3. As $9+8+7+6+5 = 35$, we must eliminate a digit, the remainder of which divided by 3 is 2, that is 8. Now, the digit sum $9+7+6+5 = 27$ is divisible by 3. The largest number formed by 9, 7, 6, and 5 is 9765, which is exactly divisible by 7. So, the number we need to find is 9765.

13. As the remainders of 92, 118, and 157 divided by $n$ are the same, $118 - 92 = 26$ and $157 - 118 = 39$ are both divisible by $n$. So, $n$ divides $39 - 26 = 13$. As 13 is a prime number and $n \neq 1$, $n = 13$.

14. Among all denominators from 2 to 40, only 25 is divisible by $5 \times 5$. So, the (smallest) common denominator is $5 \times 5 \times a$, where $5 \nmid a$. In the sum, the numerator of every fraction will become a multiple of 5 after reduction to the common denominator except $\frac{1}{25}$, so $\frac{m}{n} = \frac{5b+a}{25a}$, $m = 5b + a$. As $5 \nmid a$ and $5 \mid 5b$, $m = 5b + a$ is not a multiple of 5.

# Solution 20

# Odd Numbers and Even Numbers

1. Let the numbers of students participating in the English, Chinese, and Mathematics interest groups be $x$, $y$, and $z$, respectively. According to the problem, $x$ is an even number; $y = x + 5$, so $y$ is an odd number; $z = y + 3$, so $z$ is an even number. The total number of students in the class $x + y + z$ is the sum of two even numbers and one odd number, which is an odd number. (Alternatively, $x + y + z = x + (x + 5) + (x + 5 + 3) = 3x + 13$ is odd.)

2. $1 + 3 + 5 + 7 + 9 + 11 + 13 + 15 + 17 + 19$

$$= (1 + 19) + (3 + 17) + (5 + 15) + (7 + 13) + (9 + 11)$$

$$= 20 \times 5 = 100.$$

$1 + 3 + \cdots + (2n - 3) + (2n - 1)$

$$= \frac{1}{2}\{[1 + (2n - 1)] + [3 + (2n - 3)] + \cdots + [(2n - 1) + 1]\}$$

$$= \frac{1}{2}(2n) \times n = n^2.$$

3. Tom is initially in the river. If he takes off his shoes once and puts on his shoes once, then he is still in the water. So, if he takes off his shoes for $k$ times and puts on his shoes for $k$ times, then he is still in the water. As the total number is even, he takes off his shoes as often as he puts them on. So, Tom is now in the water.

4. Among seven consecutive odd numbers, the fourth one is their average, that is, $399 \div 7 = 57$. So, these seven numbers are 51, 53, 55, 57, 59, 61, and 63.

5. The ones digit of $2 \times 4 \times 6$ is 8, the ones digit of $4 \times 6 \times 8$ is 2, and the ones digits of $6 \times 8 \times 0$, $8 \times 0 \times 2$ and $0 \times 2 \times 4$ are all zero. Now, the product is a six-digit number, $\overline{8****2}$, whose ones digit is 2, so the ones digits of these three consecutive even numbers should be 4, 6, and 8 respectively. As $104 \times 106 \times 108 > 100 \times 100 \times 100 = 1000000 > \overline{8****2}$ and $84 \times 86 \times 88 < 90 \times 90 \times 90 = 729000 < \overline{8****2}$, these three consecutive even numbers should be 94, 96, and 98. And $94 \times 96 \times 98 = 884352$ satisfies the problem condition.

6. If two people talk on the phone once, it is considered that both of them made a call, and the sum of the numbers of calls is increased by 2. Therefore, the total number of phone calls in the world is a multiple of 2, which is an even number. For people with an even number of calls, the total number of calls is a sum of several even numbers, which is still an even number. The total number of calls made by people with an odd number of calls is the total number of calls made in the world (which is an even number) minus the total number of calls made by people with an even number of calls (which is an even number), which is also an even number. We know that the sum of an odd number of odd numbers is odd and the sum of an even number of odd numbers is even. So, the number of people with an odd number of calls must be an even number.

7. $189 = 5 \times 37 + 4$. So, $A$, $B$, $C$, and $D$ have been pulled for 38 times each, and only $E$ has been pulled for 37 times. The light will be off if it has been pulled for an even number of times and will be on if it has been pulled for an odd number of times. So, only $E$ is on at the end.

8. It can be done. We denote an upright cup as 1 and an upside down cup as $-1$. Then, the original status is 1 1 1 1.

| | | | |
|---|---|---|---|
| After first flip | $-1$ | $-1$ | $-1$ | 1 |
| After second flip | $-1$ | 1 | 1 | $-1$ |
| After third flip | 1 | 1 | $-1$ | 1 |
| After fourth flip | $-1$ | $-1$ | $-1$ | $-1$ |

9. It cannot be done. As there are five odd numbers, 1, 3, 5, 7, 9, and four even numbers, 2, 4, 6, 8. So, the result must be an odd number and it cannot be the even number 10.

10. It cannot be done. $a^2 - b^2 = (a+b)(a-b)$, and $a+b$ has the same parity as $a-b$. So, if both of them are odd, then $a^2 - b^2$ is odd; if both of them are even, then $a^2 - b^2$ is divisible by 4. But 2002 is an even number not divisible by 4. So, $a^2 - b^2 \neq 2002$, that is, $a^2 \neq 2002 + b^2$.

11. It cannot be done. We denote an upright cup as 1 and an upside down cup as $-1$. Then, after every flip, four signs have been changed. So, the product of these 9 numbers is multiplied by $(-1)^4 = 1$, which will not change. At the beginning, all cups are upright, that is, the product is $+1$. Therefore, no matter how many times we flip the cups, the product will always be $+1$ and never become $-1$. But when all cups are upside down, the product should be $-1$. So, it cannot be reached. This proof is similar to Example 8.

12. Bob cannot make it. Suppose he can, then each time he leaves a room and enters a room, the parity of the room number will change. So, the route should be "odd even odd even $\cdots$ even," and the last room number should be even so as to return to room 1. Then, this requires the same number of even room numbers and odd room numbers. But from 1 to 9, there are five odd numbers and four even numbers, which are not the same.

13. It cannot be realized. Suppose that it can be done, then as there are three straight lines from each point, there are totally $3 \times 9 = 27$ straight lines. But each straight line passes through two points, so every straight line has been computed twice in the above computation, and the number of straight lines should be even. But 27 is odd, which is contradictory.

14. Label the squares on the chessboard alternately by 0 and 1. The knight, in each step, jumps from 0 to 1 or from 1 to 0. So, in the knight tour, odd numbers and even numbers appear alternately. If the knight can jump back to the original position, then the beginning and ending squares must have the same parity. So, it should be "odd even odd even $\cdots$ odd even odd" or "even odd even odd $\cdots$ even odd even." Thus, the number of steps is even.

15. There is only one even number among 2001, 2002, and 2003. So, $a$ and $c$ cannot be both even, that is, at least one of $a$ and $c$ is odd. So, at least one of $a+1$ and $c+3$ is even. Then, $(a+1)(b+2)(c+3)$ is even.

# Solution 21

# Prime Numbers and Composite Numbers

1. Among positive integers, the smallest odd prime number is 3, the smallest odd composite number is 9, and the number that is neither a prime nor a composite number is 1. The product of these three numbers is $3 \times 9 \times 1 = 27$.

2. The even prime number is 2, the smallest prime number greater than 50 is 53, and the largest prime number within 100 is 97. The sum of these three numbers is $2 + 53 + 97 = 152$.

3. As the sum of two prime numbers is 49, one of these two prime numbers is odd and the other is even, that is, one of them is 2 and the other is $49 - 2 = 47$. So, the product of these two prime numbers is $2 \times 47 = 94$.

4. As $p_1$ and $p_2$ are two prime numbers greater than 2, both of them are odd prime numbers. Then, $p_1 + p_2$ is even and greater than 2, so it is a composite number that is divisible by 2 (an even composite number).

5. As $p^2 + 3$ is a prime number greater than 2, $p^2 + 3$ is odd and $p$ must be even. Also, as $p$ is a prime number, $p = 2$. Then, $p^3 + 3 = 2^3 + 3 = 11$ is a prime number.

6. As $p$ is a prime number greater than 3, $p$ is odd and not divisible by 3. We suppose that $p = 3k + 1$ or $3k + 2$ (where $k$ is a positive integer). If $p = 3k + 1$, then $p + 2 = (3k + 1) + 2 = 3(k + 1)$ is divisible by 3 and is greater than 3, then it is not a prime number, contrary to the problem condition. So, $p = 3k + 2$, and the remainder of $p$ divided by 3 is 2.

7.  $n_1^2 - n_2^2 - 2n_1 - 2n_2 = (n_1 + n_2)(n_1 - n_2) - 2(n_1 + n_2)$
    $$= (n_1 + n_2)(n_1 - n_2 - 2).$$

    Therefore, $19 = (n_1 + n_2)(n_1 - n_2 - 2)$. As $n_1 + n_2 > 0$, $n_1 - n_2 - 2 > 0$ (otherwise, their product cannot be positive). Also, as 19 is a prime number and $n_1 + n_2 > n_1 - n_2 - 2$, we have

    $$\begin{cases} n_1 + n_2 = 19, \\ n_1 - n_2 - 2 = 1. \end{cases}$$

    On solving it, we have $n_1 = 11$ and $n_2 = 8$.

8.  The smallest four prime numbers are 2, 3, 5, and 7. So, the smallest number with four different prime factors is $2 \times 3 \times 5 \times 7 = 210$.

9.  2000 has two different prime factors, which are 2 and 5. So, the sum is $2 + 5 = 7$.

10. As $111111 + 9 \times 10^k$ is divisible by 3 and is greater than 3, it is a composite number.

11. If $n = 6k$, then $n + 3 = 6k + 3 = 3(2k + 1)$ is divisible by 3, so it is not a prime number. If $n = 6k + 2$, then $n + 7 = 6k + 9 = 3(2k + 3)$ is divisible by 3, so it is not a prime number. If $n$ is odd, then $n + 3$ is even, so it is not a prime number. Therefore, $n = 6k + 4$, that is, the remainder of $n$ divided by 6 is 4.

12. As $n$ and $n^5$ has the same parity, $n^5 - n$ is even and divisible by 2.
    $n^5 - n = n(n^4 - 1) = n(n^2 + 1)(n^2 - 1) = n(n + 1)(n - 1)(n^2 + 1)$.
    If $n = 5k$, then $n^5 - n$ is divisible by 5. If $n = 5k + 1$, then $n - 1$ is divisible by 5 and $n^5 - n$ is divisible by 5. If $n = 5k - 1$, then $n + 1$ is divisible by 5 and $n^5 - n$ is divisible by 5. If $n = 5k + 2$, then $n^2 + 1 = (5k + 2)^2 + 1 = 25k^2 + 20k + 5 = 5(5k^2 + 4k + 1)$ is divisible by 5 and $n^5 - n$ is divisible by 5. If $n = 5k - 2$, then $n^2 + 1 = (5k - 2)^2 + 1 = 5(5k^2 - 4k + 1)$ is divisible by 5 and $n^5 - n$ is divisible by 5. Therefore, in any case, $n^5 - n$ is divisible by 5. As 2 and 5 are co-prime, $n^5 - n$ is divisible by $2 \times 5$, that is, it is divisible by 10.

13. When $n = 43k + 1$ ($k$ is a positive integer), $n^2 + n + 41 = (43k + 1)^2 + (43k + 1) + 41 = (43k)^2 + 2 \times 43k + 1 + 43k + 1 + 41 = (43k)^2 + 3 \times 43k + 43 = 43(43k^2 + 3k + 1)$ is divisible by 43 and is greater than 43. So, $n^2 + n + 41$ is a composite number. As there are infinitely many $k$, there will be infinitely many $n$, such that $n^2 + n + 41$ is a composite number and is a multiple of 43.

14. Suppose that there are only finitely many prime numbers $p_1 < p_2 < \cdots < p_n$. Consider $p_1 p_2 \cdots p_n + 1$. As the remainder of $p_1 p_2 \cdots p_n + 1$ divided by $p_1$, $p_2, \ldots, p_n$ are all 1, either $p_1 p_2 \ldots p_n + 1$ is a prime number or it has a prime factor different from $p_1$, $p_2, \ldots, p_n$. This contradicts the supposition that there are only $n$ prime numbers $p_1, p_2, \ldots, p_n$. Therefore, there are infinitely many prime numbers.

15. Among nine consecutive numbers, there are at least four even numbers. As they are all greater than 80, these even numbers cannot be prime numbers. Among any three consecutive odd numbers, at least one is divisible by 3 (if the middle one is not divisible by 3, suppose that it is $3k + 1$ or $3k - 1$, then the next one is $3k + 3$ or the previous one is $3k - 3$, which are divisible by 3). As they are greater than 80, they are not prime numbers. So, among nine consecutive integers greater than 80, the number of prime numbers $\leqslant 4$. And among 101, 102, 103, 104, 105, 106, 107, 108, and 109, there are four prime numbers 101, 103, 107, and 109. So, there are at most four prime numbers among nine consecutive integers greater than 80.

# Solution 22

# The Rule of Sum and the Rule of Product

1. $12 + 20 + 15 = 47$. There are 47 ways to choose one book from 47 books.

2. $18 + 15 = 33$. There are 33 ways to select one student from 33 students.

3. (1) There are totally $2+4+5 = 11$ balls, so there are 11 different ways to take a ball from the pocket.

   (2) By the rule of product, there are $2 \times 4 \times 5 = 40$ different ways to take the balls.

4. From point A to point C, there are two methods:

   (1) Go directly to point C, and there are 4 different ways.

   (2) First go to point B and then go to point C. By the rule of product, there are $2 \times 3 = 6$ different ways.

   Therefore, by the rule of sum, there are $4 + 6 = 10$ different ways.

5. By the rule of product, there are $3 \times 4 \times 2 = 24$ different terms.

6. $2000 = 2^4 \times 5^3$, there are $(4+1)(3+1) = 20$ positive divisors.

7. There are four choices for each of the hundreds digit, tens digit, and ones digit (as digits can be chosen repeatedly). By the rule of product, there are $4 \times 4 \times 4 = 64$ three-digit numbers formed.

8. Everyone can be in group $A$ or group $B$. Therefore, everyone has two choices. There are $2 \times 2 \times 2 \times 2 \times 2 \times 2 = 64$ different ways. But we have to eliminate two cases where all people are in group $A$ and all people are in group $B$ (as in these cases, group $B$ or group $A$ is empty, which does not meet the requirement of this problem). So, there are $64 - 2 = 62$ ways.

9. There are $8(=6+2)$ different starting stations for each ticket, and there are 7 ending stations that can be reached. Therefore, there are $8 \times 7 = 56$ different tickets. If we interchange the starting and ending stations, the price should not change. So, there are at most $56 \div 2 = 28$ different prices among 56 tickets.

10. There are four choices for the first position on the left. There are three choices for the second position on the left from the remaining three people. There are two choices for the second position on the right from the remaining two people. The last person has to stand first on the right. By the rule of product, there are $4 \times 3 \times 2 = 24$ ways.

11. (1) 2, 3, and 4 can form $3 \times 2 \times 1 = 6$ three-digit numbers without repeated digits. The digit sum of each number is $2 + 3 + 4 = 9$. So, the sum of the digit sums is $9 \times 6 = 54$.

    (2) Among six three-digit numbers without repeated digits, there are two numbers whose hundreds digit is 2, 3, or 4. So, the contribution of the hundreds digits is $2 \times (2+3+4) \times 100$. For the same reason, the contribution of the tens digits is $2 \times (2+3+4) \times 10$, and the contribution of the ones digits is $2 \times (2+3+4)$. The sum of these three-digit numbers is

    $$2 \times (2+3+4) \times (100 + 10 + 1) = 2 \times 9 \times 111 = 1998.$$

12. $2000 = 2^4 \times 5^3$. The even divisors have the form $2^\alpha \times 5^\beta$, where $\alpha$ can be 1, 2, 3, or 4, totaling four choices, and $\beta$ can be 0, 1, 2, or 3, totaling four choices. So, there are $4 \times 4 = 16$ even divisors.

    **Another Solution**    $1000 = 2^3 \times 5^3$ has $(3+1) \times (3+1) = 16$ divisors. Each divisor of 1000 multiplied by 2 is an even divisor of 2000, and each even divisor of 2000 divided by 2 is a divisor of 1000. So, there are 16 even divisors of 2000.

13. (1) The thousands digit has four choices (cannot be 0), and each of the rest digits has 5 choices. So, there are $4 \times 5 \times 5 \times 5 = 500$ four-digit numbers.

    (2) The ones digit can be 0, 2, or 4, totaling 3 choices. The thousands digit has four choices, the hundreds digit has five choices, and the tens digit has five choices: $3 \times 4 \times 5 \times 5 = 300$. There are 300 four-digit even numbers.

    (3) The thousands digit has four choices. After the thousands digit is determined, the hundreds digit has four choices. After the thousands digit and the hundreds digit are determined, the tens digit

has three choices. After the thousands digit, the hundreds digit, and the tens digit are determined, the ones digit has two choices. So, $4 \times 4 \times 3 \times 2 = 96$, that is, there are 96 four-digit numbers without repeated digits.

(4) The ones digit can be 0, 2, or 4. If the ones digit is 0, there are $4 \times 3 \times 2 = 24$ four-digit numbers without repeated digits. If the ones digit is 2, there are $3 \times 3 \times 2 = 18$ four-digit numbers without repeated digits. There are also 18 four-digit numbers without repeated digits if the ones digit is 4. So, there are totally $24 + 18 + 18 = 60$ four-digit even numbers without repeated digits.

(5) Among positive numbers without repeated digits, there are 4 one-digit numbers, $4 \times 4 = 16$ two-digit numbers, $4 \times 4 \times 3 = 48$ three-digit numbers, 96 (see (3)) four-digit numbers, and $4 \times 4 \times 3 \times 2 \times 1 = 96$ five-digit numbers. The total number is $4 + 16 + 48 + 96 + 96 = 260$.

14. There are four ways to drop each letter. So, there are totally $4 \times 4 \times 4 = 64$ ways to drop the letters.

15. There are four choices for the first person on the left. After this choice is determined, there are three choices for the second person on the left. After they are determined, there are two choices for the first person on the right, and then the choice for the second person on the right will be determined. So, there are $4 \times 3 \times 2 = 24$ ways.

16. The digit sum is divisible by 3 if the number is divisible by 3. So, the sum of the other two digits is divisible by 3. If a digit is one of 1, 4, and 7, then the other one is one of 2, 5, and 8, and vice versa. There are totally $3 \times 3 = 9$ cases. Each pair of selected digits, together with 9, can form $3 \times 2 \times 1 = 6$ three-digit numbers. So, there are totally $9 \times 6 = 54$ such three-digit numbers. If all digits are among 0, 3, 6, and 9, then there are $3 \times 4 \times 4 = 48$ such three-digit numbers (of course, it is divisible by 3), $2 \times 3 \times 3 = 18$ of which are without digit 9. So, there are $48 - 18 = 30$ of them with digit 9. Therefore, there are $54 + 30 = 84$ three-digit numbers divisible by 3 and contain the number 9.

**Another Solution** There are totally $1000 - 100 = 900$ three-digit numbers. Every three consecutive numbers contain exactly one number that is divisible by 3. So, there are $900 \div 3 = 300$ three-digit numbers that is divisible by 3. Consider those without 9 among them. Of such numbers, its hundreds digit cannot be 0 or 9, which leaves eight choices. Its tens digit cannot be 9, which leaves nine choices. The ones digit can be determined by the remainder of the sum of the first two digits divided by 3. When the remainder is 0, the ones digit can be 0, 3, or 6. When the remainder is 1, the ones digit can be 2, 5, or 8. When the remainder is 2, the ones digit can be 1, 4, or 7. So, no matter how we choose the first two digits, there are three choices for the ones digit. Therefore, there are $8 \times 9 \times 3 = 216$ three-digit numbers that are divisible by 3 and do not contain the number 9. Then, the number of three-digit numbers that are divisible by 3 and contain the number 9 is $300 - 216 = 84$.

17. $A$ has four ways. After $A$ has been colored, $B$ will have three ways. After $A$ and $B$ have been colored, $C$ has two ways. After $A$, $B$, and $C$ have been colored, $D$ has two ways. After $A$, $B$, $C$, and $D$ have been colored, $E$ has two ways. So, there are totally $4 \times 3 \times 2 \times 2 \times 2 = 96$ ways.

# Solution 23

# Number of Divisors

1.

$$d(n):$$

| ones digit \ tens digit | 0 | 1 | 2 | 3 | 4 | 5 |
|---|---|---|---|---|---|---|
| 0 | | 4 | 6 | 8 | 8 | 6 |
| 1 | 1 | 2 | 4 | 2 | 2 | 4 |
| 2 | 2 | 6 | 4 | 6 | 8 | 6 |
| 3 | 2 | 2 | 2 | 4 | 2 | 2 |
| 4 | 3 | 4 | 8 | 4 | 6 | 8 |
| 5 | 2 | 4 | 3 | 4 | 6 | 4 |
| 6 | 4 | 5 | 4 | 9 | 4 | 8 |
| 7 | 2 | 2 | 4 | 2 | 2 | 4 |
| 8 | 4 | 6 | 6 | 4 | 10 | 4 |
| 9 | 3 | 2 | 2 | 4 | 3 | 2 |

2. $105 = 3 \times 5 \times 7$, so it has $(1+1) \times (1+1) \times (1+1) = 8$ divisors.

3. $4500 = 2^2 \times 3^2 \times 5^3$, so it has $(2+1) \times (2+1) \times (3+1) = 36$ divisors.

4. $12 = 2 \times 2 \times 3$. So, $n$ has at most three different prime factors. To make $n$ smallest, these prime factors should be chosen from 2, 3, and 5, and the power of 2 should be the largest. If $n = 2^\alpha$, then $\alpha + 1 = 12$, $\alpha = 11$, and $n = 2^{11}$. If $n = 2^{\alpha_1} \cdot 3^{\alpha_2}$, then $(\alpha_1 + 1)(\alpha_2 + 1) = 12 = 6 \times 2 = 4 \times 3$, $\alpha_1 = 5$, $\alpha_2 = 1$ or $\alpha_1 = 3$, $\alpha_2 = 2$, $n = 2^5 \times 3$ or $n = 2^3 \times 3^2$. If $n = 2^{\alpha_1} \cdot 3^{\alpha_2} \cdot 5^{\alpha_3}$, then $\alpha_1 + 1 = 3$, $\alpha_2 + 1 = 2$, $\alpha_3 + 1 = 2$, $\alpha_1 = 2$, $\alpha_2 = 1$, $\alpha_3 = 1$, and $n = 2^2 \times 3 \times 5$. Among them, the smallest one is $n = 2^2 \times 3 \times 5 = 60$.

5. (1) $d(42) = d(2 \times 3 \times 7) = (1+1)(1+1)(1+1) = 8$.
   (2) The smallest number $n = 24$ can be found from the table in Exercise 1.
   (3) If $d(n) = 2$, then $n$ has only one prime factor $p$ and the power of $p$ is 1, so $n = p$, that is, $n$ is a prime number. If $d(n) = 3$, then $n$ has only one prime factor $p$ and the power of $p$ is 2. So, $n = p^2$.
6. Suppose that $n = p_1^{\alpha_1} p_2^{\alpha_2} \cdots p_k^{\alpha_k}$ is the decomposition of $n$ into prime factors. As $n$ is not a square, at least one of $\alpha_1, \alpha_2, \ldots, \alpha_k$ is odd. Therefore, at least one of $\alpha_1 + 1, \alpha_2 + 1, \ldots, \alpha_k + 1$ is even, and their product $d(n) = (\alpha_1 + 1)(\alpha_2 + 1) \cdots (\alpha_k + 1)$ is even.

   **Another Solution**  Pair $d$ with $\frac{n}{d}$. For different $d$, its pair $\frac{n}{d}$ is different. As all divisors of $n$ can be paired and no one is paired with itself ($d = \frac{n}{d}$ will imply $n = d^2$, which contradicts that $n$ is not a square number), the number of divisors of $n$ is even.

7. From the table in Exercise 1 and $d(60) = 12$ in Exercise 4, we have $n = 2, 4, 6, 12, 24, 36, 48, 60$. And $n = 1$ can also be treated as a solution. (Numbers listed above are the first eight "rich numbers"; the ninth one is 120, and the tenth one is 144.)
8. $10 = 2 \times 5$. This natural number can be $3^9$, $5^9$, $3 \times 5^4$ or $3^4 \times 5$. The largest one is $5^9 = 1953125$.
9. By Exercise 5, a number with exactly 3 divisors must be a square of a prime number. Since $11^2 = 121 > 100$, the sum is $2^2 + 3^2 + 5^2 + 7^2 = 4 + 9 + 25 + 49 = 87$.
10. Numbers with an odd number of divisors are square numbers. From 300 to 600, they are $18^2 = 324$, $19^2 = 361$, $20^2 = 400$, $21^2 = 441$, $22^2 = 484$, $23^2 = 529$, and $24^2 = 576$.
11. $240 = 2^4 \times 3 \times 5$. So, the sum of the divisors of 240 is

$$(1 + 2 + 2^2 + 2^3 + 2^4) \times (1 + 3) \times (1 + 5) = 31 \times 4 \times 6 = 744.$$

12. The number of apples taken out each time should be a divisor of 96. This problem requires simply to find the number of divisors of 96, excluding 1 and 96. $d(96) = d(2^5 \times 3) = (5+1)(1+1) = 12$, $12 - 2 = 10$. That is, there are 10 ways.

# Solution 24

# Positional Notation

1. $(102102)_3 = 1 \times 3^5 + 2 \times 3^3 + 1 \times 3^2 + 2$
   $$= 243 + 54 + 9 + 2 = 308.$$

2. 

   $$
   \begin{array}{r|rrrl}
   3 & 3 & 0 & 8 & \\
   \cline{2-4}
   3 & 1 & 0 & 2 & \cdots\cdots \quad 2 \\
   \cline{2-3}
    & 3 & 3 & 4 & \cdots\cdots \quad 0 \\
   \cline{2-3}
    & 3 & 1 & 1 & \cdots\cdots \quad 1 \\
   \cline{2-3}
    & & 3 & 3 & \cdots\cdots \quad 2 \\
   \cline{3-3}
    & & & 1 & \cdots\cdots \quad 0 \\
   \end{array}
   $$

   So, $308 = (102102)_3$.

3. We only need to convert each digit in base-8 into binary. Except for the first digit, add zeros to those with fewer than three digits: $(3)_8 = (11)_2$, $(7)_8 = (111)_2$, $(1)_8 = (001)_2$, and $(4)_8 = (100)_2$, and then connect them in order. We have

   $$(3714)_8 = (11111001100)_2.$$

4. We first convert it into decimal:

   $$(111002220)_3 = 2 \times 3 + 2 \times 3^2 + 2 \times 3^3 + 1 \times 3^6 + 1 \times 3^7 + 1 \times 3^8$$
   $$= 6 + 18 + 54 + 729 + 2187 + 6561 = 9555.$$

Then, we convert it into quinary.

$$
\begin{array}{rl|llll}
5 & \!\!| & 9 & 5 & 5 & 5 \\ \cline{3-6}
5 & \!\!| & 1 & 9 & 1 & 1 \quad \cdots\cdots \quad 0 \\ \cline{3-6}
& 5 & \!\!| & 3 & 8 & 2 \quad \cdots\cdots \quad 1 \\ \cline{4-6}
& & 5 & \!\!| & 7 & 6 \quad \cdots\cdots \quad 2 \\ \cline{5-6}
& & & 5 & \!\!| & 1 \; 5 \quad \cdots\cdots \quad 1 \\ \cline{6-6}
& & & & & 3 \quad \cdots\cdots \quad 0
\end{array}
$$

So, $(111002220)_3 = (301210)_5$.

5. (1)

$$
\begin{array}{r}
1\;1 \\
110\,\overline{)\,1\;0\;0\;1\;0} \\
1\;1\;0 \\ \hline
1\;1\;0 \\
1\;1\;0 \\ \hline
0
\end{array}
$$

$$
\begin{array}{r}
1\;1\;1 \\
11\,\overline{)\,1\;0\;1\;0\;1} \\
1\;1 \\ \hline
1\;0\;0 \\
1\;1 \\ \hline
1\;1 \\
1\;1 \\ \hline
0
\end{array}
$$

So, the original formula $= (11)_2 + (111)_2 = (1010)_2$.

(2)

$$
\begin{array}{r}
1\;0\;1\;1 \\
\times \quad\quad 1\;1 \\ \hline
1\;0\;1\;1 \\
1\;0\;1\;1 \quad\; \\ \hline
1\;0\;0\;0\;0\;1
\end{array}
$$

$$
\begin{array}{r}
1\;0\;0\;1\;0\;0 \\
+ \quad 1\;1\;0\;1\;1 \\ \hline
1\;1\;1\;1\;1\;1 \\
- \;1\;0\;0\;0\;0\;1 \\ \hline
1\;1\;1\;1\;0
\end{array}
$$

So, the original formula $= (11110)_2$.

6.

```
      1 0 0 0 1 0 0 0 1 0 0
  −     1 0 0 0 1 0 0 0 1
    ─────────────────────
      1 1 0 0 1 1 0 0 1 1

          1 0 1 1 0 1 1
  ×                 1 1
    ─────────────────────
          1 0 1 1 0 1 1
        1 0 1 1 0 1 1
    ─────────────────────
        1 0 0 0 1 0 0 0 1

                1 0 1 1 0 1 1
  1001 )   1 1 0 0 1 1 0 0 1 1
           1 0 0 1
         ───────────
               1 1 1 1
               1 0 0 1
             ───────────
               1 1 0 0
               1 0 0 1
             ───────────
                   1 1 0 1
                   1 0 0 1
                 ───────────
                     1 0 0 1
                     1 0 0 1
                   ───────────
                             0
```

So, the original formula $= (100010001)_2$.

7. Suppose the equation holds in base-$g$. As the largest digit is 4, $g \geqslant 5$. As we know that in decimal $4 \times (4 \times g + 1) = 3 \times g^2 + g + 4$, we get $16 \times g = 3 \times g^2 + g$, $3 \times g^2 = 15 \times g$, and $3 \times g = 15$, $g = 5$. That is, in quinary, $4 \times 41 = 314$.

8. Starting from the right side, make two-digit sections 22, 22, 11, 11, 22, 12, 11, 22, 11, and 12. Each section determines a digit in base-9. In the first section, $(12)_3 = 1 \times 3 + 2 = 5$. So, in base-9, the first digit from the left is 5.

9. Note that in decimal $11 \times 11 = 121 = 10 \times 12 + 1$, so $(e)_{12} \times (e)_{12} = (\overline{t1})_{12}$.

```
              e   e   e
  ×           e   e   e
  ───────────────────────
          t   e   e   1
      t   e   e   1
  t   e   e   1
  ───────────────────────
  e   e   t   0   0   1
```

That is, the square of $(\overline{eee})_{12}$ is $(eet001)_{12}$.

10. In decimal, $1111^2 = 1111 \times 1111 = 1234321$. The same happens in base-$g$ through the following multiplication.

$$
\begin{array}{rrrrrrr}
 &   &   & 1 & 1 & 1 & 1 \\
 &   & \times & 1 & 1 & 1 & 1 \\
\hline
 &   &   & 1 & 1 & 1 & 1 \\
 &   & 1 & 1 & 1 & 1 &   \\
 & 1 & 1 & 1 & 1 &   &   \\
1 & 1 & 1 & 1 &   &   &   \\
\hline
1 & 2 & 3 & 4 & 3 & 2 & 1 \\
\end{array}
$$

So, $(1234321)_g = (1111)_g^2$.

11. As $a \times 9^2 + b \times 9 + c = c \times 7^2 + b \times 7 + a$, that is, $80a + 2b = 48c$ and $40a + b = 24c$, we have $b = 24c - 40a = 8(3c - 5a)$, and $b$ is divisible by 8, but $0 \leqslant b \leqslant 6$, so $b = 0$, $3c - 5a = 0$. So, $c$ is divisible by 5 and $a$ is divisible by 3, and both of them are less than 7, not equal to 0. So, $c = 5$ and $a = 3$. $(305)_9 = 3 \times 9^2 + 5 = 248$.

12. Put $1$, $2$, $2^2$, $2^3, \ldots, 2^8$, and $2^9 - 23$ bullets into the boxes. For any number $n$ less than $512 = 2^9$, it can be represented as $a_0 \times 2^8 + a_1 \times 2^7 + \cdots + a_7 \times 2 + a_8$ (that is, $(\overline{a_0 a_1 \cdots a_8})_2$), and he can take away some boxes from the first nine boxes to get $n$ bullets. (For example, if $a_0 = 1$, he takes the box with $2^8$ bullets; if $a_0 = 0$, he does not take it. It is similar for the other boxes.) If number $n$ is between 512 and 1000, $n - (2^9 - 23) = n - 489 \leqslant 1000 - 489 = 511$. So, he can first take away the box with $2^9 - 23$ bullets and then take away some boxes from the first nine boxes so as to get $n$ bullets.

# Solution 25

# Modular Arithmetic

1. Obviously, $n = 1$ satisfies $n^2 \equiv n \pmod{30}$. Numbers of the form $1 + 30k$ ($k$ is an integer) satisfy $n^2 \equiv n \pmod{30}$. Therefore, there are infinitely many natural numbers satisfying the condition.

2. $10 \equiv 1 \pmod 9$, so for any positive integer $k$,
$$10^k \equiv 1 \pmod 9.$$

$$m = a_n \times 10^{n-1} + a_{n-1} \times 10^{n-2} + \cdots + a_2 \times 10 + a_1$$

$$\equiv a_n + a_{n-1} + \cdots + a_2 + a_1 \pmod 9.$$

3. $A = 2001^{2002} \equiv (2+1)^{2002} = 9^{1001} \equiv 0 \pmod 9$.
   By Exercise 2, we have $D \equiv C \equiv B \equiv A \equiv 0 \pmod 9$.
   Also,

$$A < (10^4)^{2002} = 10^{8008},$$

$$B < 9 \times 8008 < 100000,$$

$$C < 9 \times 5 = 45 < 100,$$

$$D < 9 \times 2 = 18.$$

   So, $D = 9$.

4. $17 \equiv 2 \pmod{15}$, $2^4 = 16 \equiv 1 \pmod{15}$, so

$$17^{2013} \equiv 2^{2013} = 2^{4 \times 503 + 1} \equiv 2 \pmod{15},$$

$$17^{2013} - 2 \equiv 0 \pmod{15}.$$

5. $10^{2013} + 23^{2015} \equiv (-1)^{2013} + 1^{2015} = -1 + 1 = 0 \pmod{11}$.

6. Consider the ones digit of $a$. It is not difficult to check $a = 0, 1, 2, \ldots, 9$ and see that the ones digits of $a^5$ and $a$ are the same, that is,

$$a^5 \equiv a \pmod{10}.$$

(For example, $2^5 = 32 \equiv 2 \pmod{10}$.)

7. In Exercise 6, we have proved that for any positive integer $a$, $a^5 \equiv a \pmod{10}$. When $a$ is a negative integer, by multiplying $-1$ on both sides of $(-a)^5 \equiv -a \pmod{10}$, we have $a^5 \equiv a \pmod{10}$. So, for any integer $a$, we have $a^5 \equiv a \pmod{10}$.

$$a^{2049} - a^{2013} \equiv a^{2045} - a^{2013} \equiv a^{2041} - a^{2013}$$

$$\equiv \cdots \equiv a^{2017} - a^{2013}$$

$$\equiv a^{2013} - a^{2013} \equiv 0 \pmod{10}.$$

8. $x \equiv 2 \equiv 5 \pmod{3}$, $x \equiv 1 \equiv 5 \pmod{4}$. That is, $x - 5$ is divisible by 3 and 4. Therefore, $x - 5$ is divisible by 12, that is, $x \equiv 5 \pmod{12}$.

9. $72 = 8 \times 9$, so $\overline{x679y}$ is divisible by 9 and 8.

$$x + 6 + 7 + 9 + y \equiv \overline{x679y} \equiv 0 \pmod{9},$$

so $x + y + 4 \equiv 0 \pmod{9}$.

$$790 + y \equiv 0 \pmod{8},$$

so $y \equiv 2 \pmod{8}$.

As $0 \leqslant y < 10$, $y = 2$;

$$x \equiv -2 - 4 \equiv 3 \pmod{9},$$

so $x = 3$.

10. $0 \equiv n^{2014} + 2006 \equiv (-1)^{2014} + 2006 = 2007 \pmod{n+1}$. The divisors of $2007 = 3 \times 3 \times 223$ greater than 1 are

$$3, 9, 223, 669, 2007.$$

So, $n = 2, 8, 222, 668, 2006$.

11. $2^{10} = 1024 \equiv -1 \pmod{25}$, so

$$2^{999} \equiv 2^9 \times (-1)^{99} \equiv -2^9 \pmod{25}.$$

Also, it is obvious that $2^{999} \equiv -2^9 \equiv 0 \pmod 4$, so

$$2^{999} \equiv -2^9 = -512 \equiv 88 \pmod{100},$$

that is, the last two digits of $2^{999}$ are 88.

12. As $1^5 + 2^5 + 3^5 + 4^5 \equiv 1 + 0 + 4 + 0 \equiv 0 \pmod 4$, so

$$1^5 + 2^5 + 3^5 + 4^5 + \cdots + 2013^5 \equiv 2013^5 \equiv 1 \pmod 4,$$

that is, the remainder is 1.

# Solution 26

# First-Degree Diophantine Equation with Two Unknowns

1. (1) As the greatest common divisor 2 of 2 and 6 does not divide 5, the equation has no integer solutions.
   (2) As the greatest common divisor 2 of 4 and 6 divides 8, the equation has integer solutions.
   (3) The equation is simply $3x - 6y = 6$. As the greatest common divisor 3 of 3 and 6 divides 6, the equation has integer solutions.
   (4) The equation is simply $3x + 2y = 11$. As the greatest common divisor 1 of 3 and 2 divides 1, the equation has integer solutions.

2. (1) $2x = 7 - 6y$, $x = \frac{7-6y}{2}$. So, the solution of the equation is
   $$\begin{cases} x = \frac{7-6k}{2}, \\ y = k, \end{cases}$$ where $k$ is any number.
   (2) $x = -\frac{4}{3}y - 3$. The solution of the equation is $\begin{cases} x = -\frac{4}{3}k - 3, \\ y = k, \end{cases}$ where $k$ is any number.

3. (1) $x + 2y = 4$. $y = \frac{4-x}{2} = 2 - \frac{x}{2}$. Set $x = 2k$, where $k$ is an integer, then $y = 2 - k$. The integer solution of the equation is $\begin{cases} x = 2k, \\ y = 2 - k, \end{cases}$ where $k$ is any integer.
   (2) $4y = 3x - 7$, $y = \frac{3x-7}{4} = \frac{3x-3}{4} - 1 = \frac{3(x-1)}{4} - 1$. Set $\frac{x-1}{4} = k$, where $k$ is an integer, then the integer solution of the equation is $\begin{cases} x = 4k + 1, \\ y = 3k - 1, \end{cases}$ where $k$ is any integer.

(3) $7y = 8 - 4x$, $y = \frac{8-4x}{7} = \frac{4(2-x)}{7}$. Set $\frac{2-x}{7} = k$, then the integer solution of the equation is $\begin{cases} x = 2 - 7k, \\ y = 4k, \end{cases}$ where $k$ is any integer.

(4) $y = \frac{4-13x}{30} = 1 - \frac{26+13x}{30} = 1 - \frac{13(2+x)}{30}$. Set $\frac{2+x}{30} = k$, then the integer solution of the equation is $\begin{cases} x = 30k - 2, \\ y = 1 - 13k, \end{cases}$ where $k$ is any integer.

4. (1) As $11 + 15 > 20$, the equation has no positive integer solutions.

(2) $5y = 21 - 2x \leqslant 21 - 2 < 20$, $y < 4$. When $y = 1$, we have $x = \frac{21-5}{2} = 8$. When $y = 2$, $x = \frac{21-10}{2}$ is not an integer. When $y = 3$, we have $x = \frac{21-15}{2} = 3$.

The positive integer solutions of the equation are $\begin{cases} x = 3, \\ y = 3, \end{cases}$ and $\begin{cases} x = 8, \\ y = 1. \end{cases}$

(3) As $x = -1$ and $y = -1$ is a solution of the original equation, all integer solutions of the equation are $\begin{cases} x = -1 + 2k, \\ y = -1 + 5k, \end{cases}$ where $k$ is an integer. By $\begin{cases} -1 + 2k > 0, \\ -1 + 5k > 0 \end{cases}$ we have $k \geqslant 1$.

So, the positive integer solutions of the equation are $\begin{cases} x = -1 + 2k, \\ y = -1 + 5k, \end{cases}$ where $k$ is a positive integer.

(4) $x = 0$ and $y = 4$ is a solution of the equation. So, the general solution of the equation is $\begin{cases} x = 8k, \\ y = 4 - 5k. \end{cases}$ Let

$$\begin{cases} 8k > 0, & (26.1) \\ 4 - 5k > 0. & (26.2) \end{cases}$$

We have $0 < k < \frac{4}{5}$. But there is no integer in this range, so the equation has no positive integer solutions.

5. Suppose that one integer is $11x$ and the other is $17x$. As $y$ is a positive integer, we find

$$11x + 17y = 100.$$

$$17y = 100 - 11x \leqslant 100 - 11 = 89, y \leqslant \frac{89}{17}.$$

As $y$ is a positive integer, $y \leqslant 5$. $x = \frac{100-17y}{11}$. It is not difficult to check that, among $y = 1, 2, 3, 4, 5$, only when $y = 2$, $x$ is an integer and $x = 6$ at this time. So, $100 = 66 + 34$.

6. Suppose that Bob works in company A for $x$ months and he works in company B for $y$ months, where $x$ and $y$ are positive integers. Then, $470x + 350y = 7620$, that is,

$$47x + 35y = 762. \tag{26.3}$$

According to the problem conditions $y < x \leqslant 12$, so we have $(47 + 35)x > 762$, $x > \frac{762}{82} > 9$. As $x$ is an integer, so $x \geqslant 10$. Substitute $x = 10, 11, 12$ into (26.3), we have that only $x = 11$ makes $y = \frac{762-47x}{35}$ an integer. So, the positive integer solution of (26.3) is $\begin{cases} x = 11, \\ y = 7. \end{cases}$

Bob works in company A for 11 months, and he works in company B for 7 months.

Remark: To find the positive integer solutions, usually, we first determine the ranges of the unknowns $x$ and $y$ and then substitute the possible values of $x$ or $y$ one by one to check whether they give positive integer solutions.

# Solution 27

# The Drawer Principle

1. There are only 12 zodiac signs. We regard them as 12 drawers and the 13 students as 13 apples. According to the drawer principle, there must be two apples in the same drawer; that is, there are two students whose zodiac signs are the same.

2. Divide 1 min into 60 s, and regard every second as a drawer; regard the 65 hits as 65 apples. Then, there must be a drawer with at least two apples; that is, there is a second in which the table tennis player hits the ball more than once.

3. As shown in Figure 27.1, we divide the equilateral triangle with side length 1 into four small equilateral triangles with side length $\frac{1}{2}$. Two of the five points must fall in the same small equilateral triangle, and the distance between them will not exceed the side length, which is $\frac{1}{2}$.

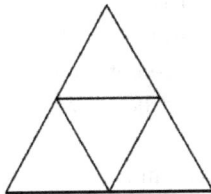

Fig. 27.1

4. When an integer is divided by 10, the remainder is one of the following 10 numbers: 0, 1, 2, 3, 4, 5, 6, 7, 8, or 9. Regard them as 10 drawers, and all natural numbers with remainder $k$ are put into the $k$th drawer $(0 \leqslant k \leqslant 9)$. Among the 11 natural numbers, there must be two in the

same drawer, and they have the same remainder. So, their difference is a multiple of 10.

5. Group integers from 1 to 100 in the following way: $(1, 1 \times 2, 1 \times 2^2, \ldots, 1 \times 2^6)$, $(3, 3 \times 2, 3 \times 2^2, \ldots, 3 \times 2^5)$, $(5, 5 \times 2, 5 \times 2^2, \ldots, 5 \times 2^4), \ldots, (49, 49 \times 2), (51), (53), \ldots, (99)$. There are totally 50 groups, and we regard them as 50 drawers. For 51 numbers, there must be two of them in the same drawer, then the larger one is a multiple of the smaller one.

6. Group these 8 numbers as $(1, 15), (3, 13), (5, 11), (7, 9)$. When we select 5 numbers, there must be two numbers in the same group, and their sum is 16.

7. Regard 6 people as 6 points. For two people who know each other, we connect the two points by a red line; otherwise, we connect them by a blue line. By Example 5, there is always a triangle in the graph such that its three sides have the same color. The corresponding three people know each other (if the three sides are red) or do not know each other (if the three sides are blue).

8. If we take out 1 red ball, 1 blue ball, and 1 white ball, totaling 3 balls, then there are no two balls of the same color. If we take out 4 balls, then as there are only three colors of the balls (3 drawers), two of the balls must be of the same color. So, at least 4 balls must be taken out to ensure that 2 of them have the same color.

9. Group integers by their ones digit (from 0 to 9) into six categories: $(0)$, $(5)$, $(1, 9)$, $(2, 8)$, $(3, 7)$, and $(4, 6)$. Among any 7 numbers, there must be two in the same category. Their sum (if they have different ones digits) or difference (if they have the same ones digit) is divisible by 10.

10. As $(11 - 1) \times 2 = 20$, so the 10 distances between adjacent potted flowers cannot be all greater than 2, and there must be two potted flowers such that the distance between them is no more than 2. On the other hand, we put 1 potted flower at one end of the balcony and put another potted flower near it. For the rest, put them one by one 2.1 m apart. In this way, only the first two potted flowers have a distance of no more than 2 m. So, at least 2 potted flowers have a distance of no more than 2 m.

11. Take out 5 blue balls and 7 white balls, totaling 12 balls. Then, one color (red) is missing. On the other hand, when we take out 13 balls, as $13 > 3 + 5$, $13 > 3 + 7$, and $13 > 5 + 7$, all three colors appear among them. Therefore, at least 13 balls must be taken

out to ensure that there are balls of all colors among the balls taken out.

12. There are 16 people discussing issues with $A$. There are three issues, denoted by yellow, red, and blue. There must be 6 people ($16 \div 3 = 5 \cdots 1$) discussing the same issue with $A$. Without loss of generality, we suppose that they are $B$, $C$, $D$, $E$, $F$, and $G$ and they discuss the yellow issue with $A$. If two of $B$, $C$, $D$, $E$, $F$, and $G$, such as $B$ and $C$, discuss the yellow issue, then $A$, $B$, and $C$ discuss the same issue between each other. If $B$, $C$, $D$, $E$, $F$, and $G$ do not discuss the yellow issue between each other, then the issues they discuss must be blue and red. Regard these 6 people as points, and connect every two points by the color of the issue they discuss. By Example 5, there is a triangle such that the colors of its three sides are the same. The corresponding three people discuss the same issue between each other.

13. Consider $n$ numbers

$$a_1, a_1 + a_2, a_1 + a_2 + a_3, \ldots, a_1 + a_2 + \cdots + a_n. \qquad (27.1)$$

If one of them is a multiple of $n$, then the conclusion is valid. If all of them are not divisible by $n$, then the remainders of them divided by $n$ are not 0. Now, there are only $n - 1$ possible remainders, that is, $1, 2, \ldots, n - 1$, and there must be 2 numbers among these $n$ numbers such that they have the same remainder, that is, their difference is divisible by $n$. As for the difference of any two numbers in (27.1), it is the sum of several consecutive terms of $a_1, a_2, \ldots, a_n$, so the conclusion is valid.

14. $21879 = 9 \times 11 \times 13 \times 17$. Among 17 numbers, either a number $a_1$ is divisible by 17 or none of the 17 numbers are divisible by 17. For the latter case, the remainders of these 17 numbers divided by 17 have 16 cases, that is, $1, 2, \ldots, 16$. Therefore, there must be two of them, $a_1$ and $a_2$, such that the remainders of them divided by 17 are the same, that is, $a_1 - a_2$ is divisible by 17. Now, consider the remaining 15 numbers. For the same reason, there must be a number $a_3$ that is divisible by 13 or two numbers $a_3$ and $a_4$ such that the difference $a_3 - a_4$ is divisible by 13. Then, among the remaining 13 numbers, there must be a number $a_5$ that is divisible by 11 or two numbers $a_5$ and $a_6$ such that the difference $a_5 - a_6$ is divisible by 11. Finally, among the rest 11 numbers, there must be a number $a_7$ that is divisible by 9 or two numbers $a_7$ and $a_8$ such that the difference $a_7 - a_8$ is divisible by 9. Take the product of the numbers (one number or the difference between

two numbers) which are divisible by 17, 13, 11, and 9, respectively, and multiply it by the remaining numbers. The result will be a multiple of 21879.

15. Group the natural numbers by their remainders divided by 100 into the following 51 groups: $(0)$, $(1, 99)$, $(2, 98)$, ..., $(49, 51)$, $(50)$. For any 52 natural numbers, there must be two in the same group. Their difference (if their remainders are the same) or sum (if their remainders are different) is divisible by 100.

16. The numbers of times of shaking hands can be $0, 1, 2, \ldots, n-1$, where 0 and $n-1$ cannot appear at the same time. If a person shakes hands 0 times, that means no one shakes hands with him, thereby everyone shakes hands $n-2$ times at most. So, there are at most $n-1$ different numbers of times of shaking hands. Then, among $n$ people, there are at least two people such that they shake hands the same number of times.

17. Regard these 2007 numbers as $a_1, a_2, \ldots, a_{2007}$. The conclusion follows from Exercise 13 ($n = 2007$).

www.ingramcontent.com/pod-product-compliance
Lightning Source LLC
Chambersburg PA
CBHW061620220326
41598CB00026BA/3820